Detective Stories
Classic

經典偵探故事

商務印書館

This Chinese edition of *Classic Detective Stories* has been published with the written permission of Black Cat Publishing.

The copyright of this Chinese edition is owned by The Commercial Press (H.K.) Ltd.

Name of Book: Classic Detective Stories
Authors: Arthur Conan Doyle, Charles Dickens,
        Clarence Rook, Gilbert Keith Chesterton
Text adaptation and notes: James Butler
Activities: Kenneth Brodey
Editors: Monika Marszewska, Rebecca Raynes
Design: Nadia Maestri
Illustrations: Didi Coppola
Edition: © 2002 Black Cat Publishing
        an imprint of Cideb Editrice, Genoa, Canterbury

系 列 名：Black Cat 優質英語階梯閱讀 · Level 4
書　　名：經典偵探故事
責任編輯：傅　伊
封面設計：張毅　曹磊
出　　版：商務印書館 (香港) 有限公司
　　　　　香港筲箕灣耀興道 3 號東滙廣場 8 樓
　　　　　http://www.commercialpress.com.hk
發　　行：香港聯合書刊物流有限公司
　　　　　香港新界大埔汀麗路 36 號中華商務印刷大廈 3 字樓
印　　刷：中華商務彩色印刷有限公司
　　　　　香港新界大埔汀麗路 36 號中華商務印刷大廈
版　　次：2012 年 7 月第 5 次印刷
　　　　　© 商務印書館 (香港) 有限公司
　　　　　ISBN 978 962 07 1650 8
　　　　　Printed in Hong Kong

出版説明

　　本館一向倡導優質閱讀，近年來連續推出了以"Q"為標識的"Quality English Learning 優質英語學習"系列，其中《讀名著學英語》叢書，更是香港書展入選好書，讀者反響令人鼓舞。推動社會閱讀風氣，推動英語經典閱讀，藉閱讀拓廣世界視野，提高英語水平，已經成為一種潮流。

　　然良好閱讀習慣的養成非一日之功，大多數初、中級程度的讀者，常視直接閱讀厚重的原著為畏途。如何給年輕的讀者提供切實的指引和幫助，如何既提供優質的學習素材，又提供名師的教學方法，是當下社會關注的重要問題。針對這種情況，本館特別延請香港名校名師，根據多年豐富的教學經驗，精選海外適合初、中級英語程度讀者的優質經典讀物，有系統地出版了這套叢書，名為《Black Cat 優質英語階梯閱讀》。

　　《Black Cat 優質英語階梯閱讀》體現了香港名校名師堅持經典學習的教學理念，以及多年行之有效的學習方法。既有經過改寫和縮寫的經典名著，又有富創意的現代作品；既有精心設計的聽、説、讀、寫綜合練習，又有豐富的歷史文化知識；既有彩色插圖、繪圖和照片，又有英美專業演員朗讀作品的 CD。適合口味不同的讀者享受閱讀之樂，欣賞經典之美。

　　《Black Cat 優質英語階梯閱讀》由淺入深，逐階提升，好像參與一個尋寶遊戲，入門並不難，但要真正尋得寶藏，需要投入，更需要堅持。只有置身其中的人，才能體味純正英語的魅力，領略得到真寶的快樂。當英語閱讀成為自己生活的一部分，英語水平的提高自然水到渠成。

商務印書館（香港）有限公司
編輯部

使用說明

1 應該怎樣選書？

按閱讀興趣選書

《Black Cat 優質英語階梯閱讀》精選世界經典作品，也包括富於創意的現代作品；既有膾炙人口的小說、戲劇，又有非小說類的文化知識讀物，品種豐富，內容多樣，適合口味不同的讀者挑選自己感興趣的書，享受閱讀的樂趣。

按英語程度選書

《Black Cat 優質英語階梯閱讀》現設 Level 1 至 Level 6，由淺入深，涵蓋初、中級英語程度。讀物分級採用了國際上通用的劃分標準，主要以詞彙（vocabulary）和結構（structures）劃分。

Level 1 至 Level 3 出現的詞彙較淺顯，相對深的核心詞彙均配上中文解釋，節省讀者查找詞典的時間，以專心理解正文內容。在註釋的幫助下，讀者若能流暢地閱讀正文內容，就不用擔心這一本書程度過深。

Level 1 至 Level 3 出現的動詞時態形式和句子結構比較簡單。動詞時態形式以現在時（present simple）、現在時進行式（present continuous）、過去時（past simple）為主，句子結構大部分是簡單句（simple sentences）。此外，還包括比較級和最高級（comparative and superlative forms）、可數和不可數名詞（countable and uncountable nouns）以及冠詞（articles）等語法知識點。

Level 4 至 Level 6 出現的動詞時態形式，以現在完成時（present perfect）、現在完成時進行式（present perfect continuous）、過去完成時（past perfect continuous）為主，句子結構大部分是複合句（compound sentences）、條件從句（1st and 2nd conditional sentences）等。此外，還包括情態動詞（modal verbs）、被動形式（passive forms）、動名詞（gerunds）、

短語動詞（phrasal verbs）等語法知識點。

　　根據上述的語法範圍，讀者可按自己實際的英語水平，如詞彙量、語法知識、理解能力、閱讀能力等自主選擇，不再受制於學校年級劃分或學歷高低的約束，完全根據個人需要選擇合適的讀物。

② 怎樣提高閱讀效果？

　　閱讀的方法主要有兩種：一是泛讀，二是精讀。兩者各有功能，適當地結合使用，相輔相成，有事半功倍之效。

　　泛讀，指閱讀大量適合自己程度（可稍淺，但不能過深）、不同內容、風格、體裁的讀物，但求明白內容大意，不用花費太多時間鑽研細節，主要作用是多接觸英語，減輕對它的生疏感，鞏固以前所學過的英語，讓腦子在潛意識中吸收詞彙用法、語法結構等。

　　精讀，指小心認真地閱讀內容精彩、組織有條理、遣詞造句又正確的作品，着重點在於理解 "準確" 及 "深入"，欣賞其精彩獨到之處。精讀時，可充分利用書中精心設計的練習，學習掌握有用的英語詞彙和語法知識。精讀後，可再花十分鐘朗讀其中一小段有趣的文字，邊唸邊細心領會文字的結構和意思。

　　《Black Cat 優質英語階梯閱讀》中的作品均值得精讀，如時間有限，不妨嘗試每兩個星期泛讀一本，輔以每星期挑選書中一章精彩的文字精讀。要學好英語，持之以恆地泛讀和精讀英文是最有效的方法。

③ 本系列的練習與測試有何功能？

　　《Black Cat 優質英語階梯閱讀》特別注重練習的設計，為讀者考慮周到，切合實用需求，學習功能強。每章後均配有訓練聽、説、讀、寫四項技能的練習，分量、難度恰到好處。

聽力練習分兩類，一是重聽故事回答問題，二是聆聽主角對話、書信朗讀、或模擬記者訪問後寫出答案，旨在以生活化的練習形式逐步提高聽力。每本書均配有 CD 提供作品朗讀，朗讀者都是專業演員，英國作品由英國演員錄音，美國作品由美國演員錄音，務求增加聆聽的真實感和感染力。多聆聽英式和美式英語兩種發音，可讓讀者熟悉二者的差異，逐漸培養分辨英美發音的能力，提高聆聽理解的準確度。此外，模仿錄音朗讀故事或模仿主人翁在戲劇中的對白，都是訓練口語能力的好方法。

閱讀理解練習形式多樣化，有縱橫字謎、配對、填空、字句重組等等，注重訓練讀者的理解、推敲和聯想等多種閱讀技能。

寫作練習尤具新意，教讀者使用網式圖示（spidergrams）記錄重點，採用問答、書信、電報、記者採訪等多樣化形式，鼓勵讀者動手寫作。

書後更設有升級測試（Exit Test）及答案，供讀者檢查學習效果。充分利用書中的練習和測試，可全面提升聽、說、讀、寫四項技能。

4 本系列還能提供甚麼幫助？

《Black Cat 優質英語階梯閱讀》提倡豐富多元的現代閱讀，巧用書中提供的資訊，有助於提升英語理解力，擴闊視野。

每本書都設有專章介紹相關的歷史文化知識，經典名著更有作者生平、社會背景等資訊。書內富有表現力的彩色插圖、繪圖和照片，使閱讀充滿趣味，部分加上如何解讀古典名畫的指導，增長見識。有的書還提供一些與主題相關的網址，比如關於不同國家的節慶源流的網址，讓讀者多利用網上資源增進知識。

CONTENTS

Arthur Conan Doyle 亞瑟・柯南・道爾   9

🎧 The Five Orange Pips
五顆柑橘籽的死亡恐嚇

PART ONE	14
ACTIVITIES	22
PART TWO	27
ACTIVITIES	33
PART THREE	37
ACTIVITIES	45

Charles Dickens 查爾斯・狄更斯   50

🎧 Hunted Down
窮追到底

PART ONE	53
ACTIVITIES	57
PART TWO	61
ACTIVITIES	68
PART THREE	71
ACTIVITIES	77

English Detective
Fiction   80
英國偵探小說

Clarence Rook 克拉倫斯・魯克 84

🎧 The Stir Outside the Café Royal

小偷智擒大賊   ACTIVITIES 88
　　　　　　　　　　　　　97

Gilbert Keith Chesterton 吉爾伯特・凱思・切斯特頓 103

The Oracle of the Dog

狗的啟示   PART ONE   106

ACTIVITIES 111

PART TWO   116

ACTIVITIES 123

PART THREE 126

ACTIVITIES 133

APPENDICES

Exit Test   138
升級測試

**Key to the Activities
and Exit Test**   148
練習答案和測試答案

(PET) Cambridge Preliminary English Test-style activities

Part of the stories are recorded.　故事選錄

 These symbols indicate the beginning and end of the extracts linked to the listening activities.　聽力練習開始和結束的標記

Sir Arthur Conan Doyle,
National Portrait Gallery,
London.

$\mathscr{A}$bout the $\mathscr{A}$uthor

Arthur Conan Doyle (1859-1930) trained as a doctor. He practised medicine [1] at Southsea from 1882 to 1890. Doyle is best known for the creation of the legendary detective, Sherlock Holmes. The Sherlock Holmes stories were immensely popular with the public. When Doyle 'killed' the detective at the end of one story, he was forced to bring the character back to life.

Arthur Conan Doyle also wrote romances, books about public affairs, and a history of spiritualism. [2] He practised as an amateur [3] detective himself and there was a considerable [4] demand for his services.

The most famous Sherlock Holmes stories include:

The Adventures of Sherlock Holmes, 1892

The Memoirs of Sherlock Holmes, 1894

The Hound of the Baskervilles, 1902

---

1. **practised medicine** : was a doctor.
2. **spiritualism** : belief in the possibility of receiving messages from the spirits of the dead.
3. **amateur** : doing something for pleasure.
4. **considerable** : great in amount or size.

The Five Orange Pips [1]

INTRODUCTION

This story starts off, like many Sherlock Holmes adventures, with the great detective and his friend, Dr Watson, sitting in the Baker Street flat. The tale begins with the arrival of a mysterious visitor who has an extraordinary story to tell.

Sherlock Holmes often battles [2] against criminal organisations, as in *The Sign of Four* and *The Red-Headed League*. Here is another example of the great detective in such a struggle, although in this case the organisation is an American one. The story is also a little unusual because Holmes is defeated – not by the enemy, but by the weather!

---

1. **pips** : the small seeds of some fruit.
2. **battles** : fights.

Before you read

1 Listen to the beginning of Part One and look at the pictures. Choose the correct picture and put a tick (✓) in the box below it.

1. For what years does Dr Watson have a complete record of Holmes' cases?

A ☐

B ☐

C ☐

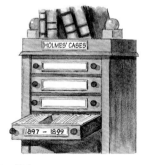

D ☐

2. When did the case of The Five Orange Pips begin?

A ☐   B ☐   C ☐   D ☐

11

3. What was the weather like on the day that the case began?

A ☐   B ☐   C ☐   D ☐

4. What was this new client like?

A ☐   B ☐

C ☐   D ☐

5. How did Holmes know that his client had come to London from the Southwest?

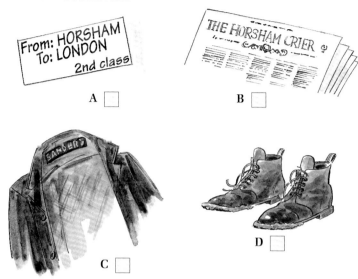

A ☐

B ☐

C ☐

D ☐

6. At what game was Major Prendergast accused of cheating?

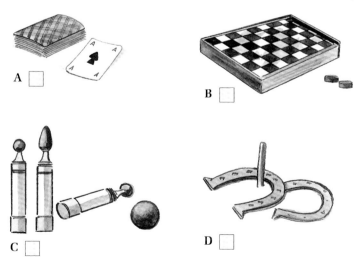

A ☐

B ☐

C ☐

D ☐

PART ONE

 have a complete record of all Sherlock Holmes'
cases between 1880 and 1897. My friend and I
worked together on some very important crimes
during that period. We also worked on some very
strange cases together. The strangest of all the cases is the
one I am going to write about now.

It all began in September. The weather was terrible, I
remember. It rained and it was very windy all day. The
weather grew worse in the evening. Sherlock Holmes and I
sat by the fire in his flat in Baker Street. We did not say
much. Holmes was working with some papers and I was
reading a story. Suddenly I heard the bell.

'I wonder [1] who that is?' I said. 'Are you expecting a
friend, Holmes?'

---

1. **wonder** : ask myself.

'No,' he said quietly. 'You're my only friend, Watson. I don't like people visiting me at home.'

'Then it must be a client,' I suggested.

'If it is a client,' Holmes replied gravely, [1] 'it is a serious case. No one would walk through this storm if the case were not serious.'

The landlady [2] opened the front door of the house. A few moments later there was a knock on the door of Holmes' flat.

'Come in!' cried Holmes.

A young man entered the room. He looked about 22 years old and he was well dressed. He seemed very nervous and he was pale.

'Give me your coat and umbrella,' Holmes ordered. 'I will hang them up to dry. I see you have come to London from the Southwest,' he added.

'Yes,' the young man agreed. He looked surprised. 'I've just come from Horsham. But how did you know that?'

'The clay and chalk [3] on your shoes is very distinctive,' [4] Holmes told him.

'I've come for advice,' said the young man.

'Advice is easy to give,' Holmes replied.

'I need help as well as advice,' the young man added.

'Help is not always easy to give,' Holmes said seriously.

'I've heard a lot about you, Mr Holmes,' the young man said. 'Major Prendergast told me how you helped him in

---

1. **gravely** : seriously.
2. **landlady** : owner of the building.
3. **clay and chalk** : soil, earth.

4. **distinctive** : having a special feature which makes something different from others.

the Tankerville Club Scandal.' [1]

'Ah, yes,' Holmes remembered with a smile. 'The Major was accused of cheating [2] at cards.'

'He said you could solve any mystery!' the young man cried.

'That was an exaggeration,' [3] Holmes said quietly.

'The Major said you are always successful!'

'That's not true,' Holmes corrected him. 'I have lost four times – three times against men and once against a woman.'

'But you've had hundreds of cases,' the young man went on. [4] 'Four defeats are nothing against hundreds of successes! I'm sure you'll be successful with my case.'

'Please tell us all about it,' my friend suggested.

'It's a strange case,' the young man began. 'The things that have happened in my family are very mysterious.'

'Tell us everything,' Holmes repeated.

'My name is John Openshaw,' the young man said. 'I have very little to do with the story. To understand it, you will have to know something about the history of my family.' He paused for a moment, then he went on. 'My grandfather had two sons – my uncle Elias and my father Joseph. My father had a bicycle factory in Coventry. He was very successful and when he retired he was a rich man.

'My uncle Elias went to America when he was a young man. He, too, became a successful man. He owned property [5]

---

1. **scandal** : a public feeling that something is improper or shocking.
2. **cheating** : playing dishonestly.
3. **exaggeration** : making something seem more than it really is.
4. **went on** : continued.
5. **property** : a piece of land, including a house.

in Florida. He fought for the South in the American Civil War. He became a Colonel in the Confederate [1] army. He did not want black people in America to have the vote. When the South was defeated, my uncle Elias returned to his property in Florida. He came back to England some years ago.

'He bought a house in Horsham. He was an odd [2] man. He was not very friendly and he lived by himself. His neighbours sometimes saw him in his garden, but he generally stayed in the house. He drank a lot of brandy and he never had any visitors. He did not want to see his brother.

'He seemed fond of me, [3] however,' Mr Openshaw continued. 'He asked my father if I could live with him. I first went to his house when I was about twelve years old. He was kind, in his own way. He played draughts [4] with me, and he put me in charge of [5] the servants in the house. By the time I was sixteen, I was master of the house. I had all the keys of the house and I could do what I wanted.

'There was only one place I couldn't go into,' Mr Openshaw said. 'There was a room in the attic that my uncle kept locked all the time. He did not allow anyone to go in there. I looked through the keyhole [6] of that room when I was a boy, but it

---

1. **Confederate** : the Confederates were the Southern States in the American Civil War. They fought to defend slavery and were defeated.

2. **odd** : strange.

3. **seemed fond of me** : seemed to like me.

4. **draughts** :

5. **put me in charge of** : made me responsible for.

6. **keyhole** :

wasn't very interesting. I could only see pieces of old luggage [1] and boxes of papers.

'One day my uncle received a letter. He looked carefully at the foreign stamp on the envelope. "From India! I wonder what it can be," he muttered. [2] He opened the letter quickly. Five orange pips fell out of it onto the table. My uncle went very pale. He looked terrified. He stared at the envelope. "KKK!" he cried loudly. He looked at the postmark on the envelope. "From Pondicherry," he said.

'"What's the matter, [3] Uncle?" I cried.

'"Death," he said. "That's what this letter means. I have done bad things in the past – and now I'm going to die!" He got up from the table and went into his room. He was still very pale. I picked up the envelope and saw the letters 'KKK' written on the inside of the flap. [4] There was no letter inside it. Just the five orange pips. I couldn't understand what was happening. I left the dining room a few minutes later and went upstairs. I saw my uncle coming down the stairs. He was carrying a key in one hand and a box in the other. He had been into the locked room in the attic.

'"They can try if they want," he muttered mysteriously. '"But I'll beat [5] them in the end." Then he spoke to me. "Call

---

1. **luggage** : suitcases and travelling bags.
2. **muttered** : said very quietly.
3. **matter** : problem.
4. **flap** :
5. **beat** : defeat.

Mr Fordham, my lawyer," he ordered.

'That afternoon the lawyer arrived. My uncle called me into
the room. There was a fire burning in the room. There were
lots of papers burning in the fire. The box from the attic room
was open on the table. I saw the letters 'KKK' on the inside of
the lid. [1]

'"I'm making a will," [2] Uncle Elias told me. "I'm leaving
everything to your father. When he dies, you will have it all,
John. Enjoy it if you can," he told me. Then he said a very odd
thing. "But if you can't enjoy it, give everything to your worst
enemy!"

'My uncle changed after that day. He began to drink a lot
more. He spent most of the time in his room. Once or twice he
came out of the room carrying a revolver. [3] He sometimes rushed
into the garden, crying that he was not afraid of anyone.

'One day he rushed into the garden with his revolver. This
time he did not come back. We found him lying at the edge of a
pond [4] in the garden. His head was in the water. He was dead.

'There was an investigation, of course. The coroner [5]
decided that Uncle Elias had committed suicide. My father
inherited [6] the property.'

---

1. **lid** : top part of a box.
2. **will** : a legal document that says who will have your money and property when you
    die.
3. **revolver** : a type of small gun.
4. **pond** : pool.
5. **coroner** : an official who inquires into the cause of a person's death when the death is
    not the result of natural causes.
6. **inherited** : received the property through the dead man's will.

Go back to the text

1 **Answer the following questions.**

   a. On what side did the young man's uncle fight during the American Civil War?

   ....

   b. What did Mr Openshaw see in his uncle's attic?

   ....

   c. What was in the envelope that Uncle Elias received?

   ....

   d. Where did the letter come from?

   ....

   e. What did the letter mean to Uncle Elias?

   ....

   f. What was written on the inside of the flap?

   ....

   g. What did Uncle Elias burn in the fire?

   ....

   h. What was written on the inside of the lid of the box?

   ....

   i. Who did Uncle Elias leave all his possessions to in his will?

   ....

   j. Where was Uncle Elias' body found?

   ....

   k. What did the coroner say about Uncle Elias' death?

   ....

PET

The case of the missing prepositions

2 For each question, choose the correct word (A, B, C or D) for each space. '0' means that no preposition is needed.

Example: I have all of my uncle's letters ...*from*.. 1870 to 1876.

    (A) from   **B** to

    C among   **D** between

1. Holmes did not have any friends except Watson because he liked living ..., ... himself.

    A with   **B** about

    C 0   **D** by

2. Just then somebody knocked .... the door.

    A by   **B** about

    C in   **D** on

3. Mr Openshaw entered .... Mr Holmes' drawing-room.

    A in   **B** 0

    C into   **D** about

4. Mr Openshaw's father told him .... the bicycle factory.

    A about   **B** from

    C by   **D** to

5. I first met Mr Holmes .... October.

    A in   **B** at

    C on   **D** by

6. Mr Openshaw went .... his uncle's study.

    A in   **B** 0

    C into   **D** about

The case of the missing letters

3 Find the words that belong to these six categories in the word square. To help you, the words are already given in the boxes but with missing letters.

LAW	GAMES	WEATHER	FAMILY
_ _ i m _	_ r _ _ g _ t _	_ t _ _ _	_ r _ _ _ _ a _ _ _ r
_ o r _ _ _ _	_ _ rd _	_ _ _ d _	_ _ c l _
_ i_ l			

WORK	HOUSES
_ a c t _ _ _	p _ _ p _ _ _ _
_ _ _ i r e _	_ _ y h _ _ _
_ l _ _ _ t	_ _ _ d l _ _ _
	_ t t _ _
	_ o o _

```
S  O  O  P  G  R  A  N  D  F  A  T  H  E  R  R
T  H  I  S  T  F  O  X  Q  U  T  A  R  Y  O  U
P  D  A  N  C  Q  R  B  Z  B  T  N  N  M  L  Q
A  R  C  Q  C  A  R  D  S  L  I  W  V  L  D  E
X  A  O  W  L  M  P  L  D  A  C  R  I  A  U  U
V  U  R  P  I  D  P  O  E  W  A  E  U  N  R  N
O  G  O  B  E  L  K  D  F  Y  E  T  U  D  D  C
L  H  N  O  N  R  L  K  A  E  L  I  P  L  L  L
A  T  E  X  T  F  T  E  C  R  M  R  S  A  A  E
T  S  R  L  I  P  C  Y  T  V  E  E  T  D  O  P
R  B  I  N  M  Y  U  H  O  L  P  D  A  Y  L  W
Q  U  A  C  P  D  O  O  R  B  R  O  I  C  L  A
S  S  T  O  R  M  I  L  Y  O  U  W  I  N  D  Y
Q  Z  B  C  R  I  M  E  O  U  Q  B  A  R  E  U
```

Before you read

PET

1 **Listen to the beginning of Part Two and choose the correct answer (A, B, C or D).**

1. When did Mr Openshaw's uncle receive the five orange pips?

   A ☐ On 2nd May 1883.

   B ☐ On 2nd May 1884.

   C ☐ On 10th March 1883.

   D ☐ On 23rd May 1873.

2. When did Mr Openshaw's uncle die?

   A ☐ On 2nd March 1883.

   B ☐ On 2nd May 1883.

   C ☐ On 22nd May 1883.

   D ☐ On 3rd March 1883.

3. What was written on the label on the inside of the box?

   A ☐ 'KKK'.

   B ☐ Civil War.

   C ☐ Records.

   D ☐ Military career.

4. What was in the box?

   A ☐ Nothing.

   B ☐ Letters, papers and receipts.

   C ☐ Records.

   D ☐ Photographs.

5. What did Mr Openshaw's uncle think about the political situation in Florida after the war?

A ☐ He was happy about the new freedom that black people had.

B ☐ He liked the politicians from the North who came to Florida.

C ☐ He had no real opinion.

D ☐ He didn't like the new freedom that black people had and he didn't like the politicians from the North.

6. How did Mr Openshaw's father feel when he received the five orange pips?

A ☐ Scared.

B ☐ Interested.

C ☐ Confused.

D ☐ Happy.

7. Where were the papers?

A ☐ On the floor in the attic.

B ☐ On the sundial.

C ☐ Uncle Elias had destroyed them.

D ☐ In the box.

8. Where was the letter posted?

A ☐ In London.

B ☐ In Pondicherry.

C ☐ In Dundee.

D ☐ In Horsham.

PART TWO

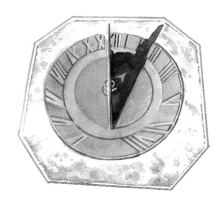

O ne moment,' said Holmes eagerly. [1] 'This is a very interesting story. I want to be sure of the facts. When did your uncle receive the letter with the five orange pips?'

'The letter arrived on the 10th of March, 1883,' Mr Openshaw answered.

'And when did he die?' Holmes asked him.

'He died seven weeks later, on the 2nd of May,' Mr Openshaw replied.

'I see,' Holmes said quietly. 'Now please go on with the story. Tell us what happened next.'

'My father examined the property very carefully,' Mr Openshaw said. 'He searched the room in the attic. The box was there. A label [2] on the inside of the box had the letters

---

1. **eagerly** : with great interest.
2. **label** : a piece of paper attached to an object that identifies it.

'KKK' written on it. There was a note on the label, which said, 'Letters, papers, receipts'. [1] The box was empty, but my father found some other papers in the attic. These were records [2] of my uncle's military career. Other papers came from the period after the Civil War. They showed that my uncle did not like the new political situation in America. He did not like the new freedom that black people had. He did not like the new politicians from the North who came to Florida.

'My father came to live in the house in Horsham at the beginning of 1884. Everything went well for about a year. Then, one morning at breakfast, he suddenly gave a cry of surprise. I looked up, and he was sitting with an envelope in one hand. In his other hand he was holding five orange pips! Of course he knew the story of the five orange pips, but he had always laughed at it. Now he looked worried.

'"What does this mean, John?" he asked me. His voice sounded scared.

'"It's 'KKK'," I replied.

He looked inside the envelope.

'"You're right," he said. "But what about this?" he asked anxiously. "What does this mean?"

'He showed me the envelope. Above the letters 'KKK' there was some writing.

'"Put the papers on the sundial [3] in the garden," I read.

---

1. **receipts :** pieces of paper proving that money has been paid.
2. **records :** *(here)* documents.
3. **sundial :**

'"What papers? I don't understand any of this."

'"The papers must be the ones from the attic," I told him. "Uncle Elias destroyed them all before he died."

'My father was worried, but he was determined to fight his fear.

'"This is all nonsense," he decided. "Where does this letter come from?"

'I looked at the postmark on the outside of the envelope. END "Dundee," I told him. "The letter was posted in Dundee."

'We were silent for a moment.

'"I think you should tell the police," I warned [1] my father.

'"They'd laugh at me!" he said quickly. "This is just a foolish [2] joke, John. We'll say no more about it."

'I tried to persuade my father to do something about the letter and the five orange pips. It was no good. He refused to do anything.

'About three days later he went to stay with an old friend of his, Major Freebody. I was glad my father was away from the house. I thought he was out of danger – but I was wrong!

'The Major sent me a telegram two days after my father's arrival. Something terrible had happened. My father had fallen over the edge of a chalk-pit [3] while he was out walking one evening. He died a few days later.

---

1. **warned** : informed of a possible danger.
2. **foolish** : stupid.
3. **chalk-pit** : open mine.

'I investigated the accident very carefully, Mr Holmes. There was no evidence of murder. The coroner decided that my father had died as a result of an accident.

'That is the story of my family,' Mr Openshaw said. 'That is how I became the owner of my uncle's house about three years ago. I have lived there very happily, Mr Holmes.'

Mr Openshaw stopped talking for a moment. He put his hand in his pocket and took out an envelope.

'Until yesterday morning that is,' he said slowly. He emptied the contents of the envelope onto the table in front of him. Five orange pips rolled out of it.

'The envelope was posted in London,' Mr Openshaw told us. 'There was the same message that my father received: "'KKK'. Put the papers on the sundial."

'What have you done about it?' Holmes wanted to know.

'Nothing,' the young man replied.

'Nothing?' Holmes repeated in surprise.

'What could I do?' Mr Openshaw asked him. 'I feel desperate [1] – like an animal in a trap!'

'You must act!' [2] Holmes announced. 'You must save yourself.'

'I went to the police,' Mr Openshaw said. 'It was no good. They listened to my story, but they didn't believe me. They just sent a policeman to the house,' he added.

'Why did you come to me?' Holmes wanted to know. 'And why didn't you come sooner?'

'I only spoke to Major Prendergast today,' the young man

---

1. **desperate** : having no hope.   2. **act** : (here) do something.

said.

Holmes began to speak quickly.

'You received the letter yesterday,' he said. 'Do you have any other evidence to show me?'

'Only this,' Mr Openshaw told him. He put a piece of blue paper on the table.

'I found this piece of paper in my uncle's room after he burnt the papers from the box,' he explained. 'It was on the floor. It seems to be a page from a diary.'

Holmes and I looked at the piece of paper. It was dated 'March, 1869', and beneath [1] it was written:

> *4th. Hudson came. Same old platform.*
> *7th. Sent the pips to McCauley,*
> *Paramore, and John Swain of*
> *St. Augustine.*
> *9th. McCauley cleared.*
> *10th. John Swain cleared.*
> *12th. Visited Paramore. All well.*

Holmes studied the piece of paper for a few minutes and then he turned to Mr Openshaw.

'You must go home at once,' he ordered him. 'Put this piece of paper into the box from the room in the attic. Then put the

---

1. **beneath** : below, under.

box on the sundial in the garden. You must also write a note. Explain that your uncle burnt all the other papers. You can do nothing else at the moment. Do you understand?'

'Yes, I do,' Mr Openshaw said. 'I'll do what you advise, Mr Holmes.'

'Go home straight away,' [1] Holmes told him. 'And be very careful – you are in great danger!'

'I'm carrying a revolver,' Mr Openshaw replied.

'Good,' Holmes replied. 'I will begin working on the case tomorrow.'

'You'll come to the house in Horsham, then?' Mr Openshaw asked him.

'No,' Holmes said. 'The secret of the case is here in London. I shall stay here to solve the mystery.'

---

1. **straight away** : immediately.

Go back to the text

1 Say whether the following statements are true (T) or false (F) and then correct the false ones.

<table>
<tr><td></td><td></td><td>**T**</td><td>**F**</td></tr>
<tr><td>**a.**</td><td>After receiving the envelope with the five orange pips, Mr Openshaw's father called the police immediately.</td><td>☐</td><td>☐</td></tr>
<tr><td>**b.**</td><td>Mr Openshaw's father went to see an old friend, Major Freebody, five days after receiving the pips.</td><td>☐</td><td>☐</td></tr>
<tr><td>**c.**</td><td>Mr Openshaw's father died five days after receiving the pips.</td><td>☐</td><td>☐</td></tr>
<tr><td>**d.**</td><td>The coroner said that somebody had pushed him into a chalk-pit.</td><td>☐</td><td>☐</td></tr>
<tr><td>**e.**</td><td>Mr Openshaw became the owner of his uncle's house after his father died.</td><td>☐</td><td>☐</td></tr>
<tr><td>**f.**</td><td>Mr Openshaw had always been very frightened living in that house.</td><td>☐</td><td>☐</td></tr>
<tr><td>**g.**</td><td>Mr Openshaw received the pips the day before he came to see Holmes.</td><td>☐</td><td>☐</td></tr>
<tr><td>**h.**</td><td>He did not come to see Holmes immediately because he did not know him until the day he came to see him.</td><td>☐</td><td>☐</td></tr>
</table>

Opposites

2 Find the opposites of the following words in the text from the list (a-i).

1. ☐ reluctantly
2. ☐ carelessly
3. ☐ full
4. ☐ intelligent
5. ☐ agreed

6. ☐ sadly
7. ☐ later
8. ☐ slowly
9. ☐ safety

a. foolish
b. sooner
c. carefully
d. danger
e. refused

f. eagerly
g. quickly
h. empty
i. happily

Now fill in the gaps in the following sentences with the opposites you have found.

a. Mr Openshaw did not go to see Mr Holmes .... because he did not know about him.

b. Mr Openshaw looked .... for the papers everywhere in the house but he couldn't find them.

c. Despite the death of his uncle and father in the house, Mr Openshaw lived there ...., until he too received a letter with the five pips.

d. Holmes told Mr Openshaw to be very careful because he knew that he was in great ....

e. Mr Holmes never .... an interesting case.

f. Mr Holmes was always bored when he wasn't working. But when a client came with an interesting case, he always listened ....

g. Mr Openshaw's father didn't want to call the police because he was sure that they would think his story was

   ....

h. The box with 'KKK' written on the inside was .... because Mr Openshaw's uncle had burned all the papers.

i. I am reading the story of *The Five Orange Pips* very .... because I really want to see how Holmes will solve the mystery.

PET

My dearest Julia!

3 Mr Openshaw has a fiancée named Julia and now that he is in great danger, he wants to explain his situation to her. 'After all,' he thinks, 'if they kill me, I want her to know my story.'

Pretend you are Mr Openshaw. You are writing this very important letter to Julia.

Tell her:

- that some people want to kill you.
- about your uncle, his life and beliefs.
- how your father died.
- about how you met Holmes and why you trust him.

Finish the letter using about 100 words.

> My dearest Julia!
>
> Something terrible has happened and I feel that I should explain everything to you. After all, we are going to be married soon and you should know everything about your future husband.
>
> It all began ....
> ....
> ....
> ....
> ....
> ....
> ....
> ....
> ....

A C T I V I T I E S

Before you read

1 Fill in the gaps in the text with the words given below. Then listen and check your answers.

few	frightened	still	kind	last
ship	died	reply	mean	after	something
explain	letter	about	leave	smoked

Mr Openshaw left the flat a little while later. Holmes and I sat in silence for a while. Then he lit his pipe and .... for a .... minutes.

'This is a strange case, Watson,' he said at last. 'John Openshaw is in very great danger – very great danger indeed!'

'What .... of danger, Holmes?' I asked excitedly.

Holmes did not .... to my question.

'Pass me the American Encyclopaedia,'[1] he said. 'I think we shall find out .... useful if we study the volume for the .... "K",' he told me. 'We also have to think .... Colonel Openshaw,' he said. 'Why did he .... America, I wonder? Was he .... of something? And why did he lead such a solitary life when he arrived here in England? Was he .... afraid of something?' He paused for a moment. 'What do the envelopes tell us?' he asked me. 'Where were the letters sent from, Watson?'

'They were sent from Pondicherry, Dundee and London,' I said.

'The .... one came from East London,' he said. 'What does that information tell you, Watson?'

'They are all seaports!' I cried excitedly. 'The writer was on a ....'

'Precisely!'[2] agreed Holmes. 'Now think about this. Colonel Openshaw .... seven weeks .... he received the orange pips. His brother died only a few days after he received the pips. How do you .... that, Watson?'

'I can't,' I admitted. 'What does it .... , Holmes?'

---

1. **encyclopaedia** : a book or set of books that gives information about every branch of knowledge.
2. **precisely** : exactly.

PART THREE

Mr Openshaw left the flat a little while later. Holmes and I sat in silence for a while. Then he lit his pipe and smoked for a few minutes. 'This is a strange case, Watson,' he said at last. 'John Openshaw is in very great danger – very great danger indeed!'

'What kind of danger, Holmes?' I asked excitedly.

Holmes did not reply to my question.

'Pass me the American Encyclopaedia,' he said. 'I think we shall find out something useful if we study the volume for the letter "K",' he told me. 'We also have to think about Colonel Openshaw,' he said. 'Why did he leave America, I wonder? Was he frightened of something? And why did he lead such a solitary life when he arrived here in England? Was he still afraid of something?' He paused for a moment. 'What do the envelopes tell us?' he asked me. 'Where were the letters sent from, Watson?'

'They were sent from Pondicherry, Dundee and London,' I said.

'The last one came from East London,' he said. 'What does that information tell you, Watson?'

'They are all seaports!' I cried excitedly. 'The writer was on a ship.'

'Precisely!' agreed Holmes. 'Now think about this. Colonel Openshaw died seven weeks after he received the orange pips. His brother died only a few days after he received the pips. How do you explain that, Watson?'

'I can't,' I admitted. 'What does it mean, Holmes?'

'The writer sends each letter on the mail boat,' [1] Holmes said. 'He then takes another boat to come to England. There is always a delay between the arrival of the letter and the death. The reason for the delay is clear. The mail boat is a fast steam vessel. [2] The writer of the letters travels on a slower boat – a sailing-ship!'

'But why, Holmes,' I asked. 'What is the reason for these murders?'

'Colonel Openshaw's papers were very important to the writer of these letters,' Holmes said. 'I think there is more than one man, Watson. There have been two murders. That suggests an organisation. 'KKK' are not the initials [3] of an individual. They are the sign of an organisation, you see. The organisation wants

---

1. **mail boat** : ship that carries post.
2. **steam vessel** : ship powered by steam, not sails.
3. **initials** : the first letters of names.

Colonel Openshaw's papers. And they will kill to get them.'

'What organisation, Holmes?'

Holmes turned the pages of the American Encyclopaedia.

'The Ku Klux Klan, Watson. It's a secret organisation that came into existence after the American Civil War. It had centres in Tennessee, Louisiana, Georgia and Florida. Colonel Openshaw lived in Florida, you remember. The purpose of the Ku Klux Klan was terrible. They were against giving black Americans the right to vote. They were very dangerous. They also had a strange tradition, Watson. If they wanted to kill a man, they sent him a warning first. They used oak leaves, melon seeds or orange pips as the warning. The victim then had a chance to change his ways, [1] or to leave the country. The Ku Klux Klan collapsed [2] in 1869.'

Holmes looked at me closely. [3]

'Openshaw came to England in 1869,' he reminded me. 'I think he was carrying the Ku Klux Klan's papers. That may be the reason for the organisation's sudden collapse. His diary contains details about the organisation's members. They are not safe until they have the diary back.'

'What about the page from the diary?' I asked. 'What does that mean?'

'It's pretty [4] clear what it means,' Holmes said. *"Sent the pips to McCauley, Paramore, and John Swain of St. Augustine."*

---

1. **change his ways** : behave differently.
2. **collapsed** : finished, disappeared.
3. **closely** : with great attention.
4. **pretty** : quite.

That's the warning, you see. The next entry [1] says, *"McCauley cleared."* That means he ran away. Then there's the final entry, *"Visited Paramore."* I expect the visit was a fatal one.'

The next morning Holmes and I had breakfast together at his flat.

'I'm worried about Mr Openshaw,' he told me. 'I may go to Horsham, after all.'

As he spoke, I picked up the newspaper that was lying on the table. I saw the headline [2] immediately.

'Holmes,' I cried, 'you're too late!'

'What do you mean?' Holmes asked quickly.

I passed him the morning newspaper.

TRAGEDY NEAR WATERLOO BRIDGE

Police Constable [3] Hook was on duty yesterday evening near Waterloo Bridge. He heard a cry for help and then a splash in the water. It was a very dark night and the weather was bad. The constable could not rescue [4] the man.

The water police found the body of a young man in the river. The man was John Openshaw of Horsham. Police believe that he was hurrying through the dark streets and fell into the river by accident. There was no sign of violence on the body.

Holmes put the newspaper down. I have never seen him look so angry.

---

1. **entry** : one item that is written down in a diary.
2. **headline** : the title of a newspaper article printed in large letters above the story.
3. **constable** : policeman of the lowest rank.
4. **rescue** : save.

'I'll get [1] them Watson. I'll find the men who did this!' my friend said. 'Openshaw came to me for help. Now he's dead.'

He thought for a moment and then he made a decision.

'I'm going out!' he announced.

'To the police?' I asked him. 'Are you going to talk to them?'

'Not yet, Watson – not until I've solved the mystery.'

I did not see my friend for the rest of the day. I returned to the flat in Baker Street early that evening. Holmes was not there so I waited for him. He came in at about 10 o'clock. He was pale and he looked very tired. He ate a piece of bread hungrily and took a long drink of water.

'You're hungry,' I commented. [2]

'I haven't eaten since this morning,' he told me. 'I've been very busy all day.'

He faced [3] me excitedly.

'I've got them, Watson. I've got them!' he cried. 'I know who they are now. And I know what I'm going to do!'

He took an orange from the table and began to pull the pips out of it. He put five pips into an envelope and wrote a name and address on it: 'Captain James Calhoun, Barque *Lone Star*, Savannah, Georgia.'

'That message will be waiting for him when he arrives,' Holmes said with a smile.

---

1. **get** : find.
2. **commented** : gave an opinion.
3. **faced** : looked at.

'But who is he? Who is this Captain Calhoun?' I asked.

'He's the leader of the organisation,' Holmes told me.

'How did you find out about him?' I asked.

Holmes smiled at me.

'I spent the day studying old newspapers,' he informed [1] me. 'I made a list of all the sailing ships that stopped at Pondicherry in January and February 1883. There were thirty-six of them. One of them was called the *Lone Star*. The name gave me a connection with America, you see.'

'Texas is sometimes called the Lone Star State,' I confirmed. [2] 'Then what did you do, Holmes?'

'I made a list of all the sailing ships that stopped in Dundee in January 1885,' Holmes said. 'Again, the *Lone Star* was one of them. Then I discovered that the *Lone Star* arrived in London a week ago. She has left London now and is returning to Savannah.'

'What are you going to do?'

---

1. **informed** : told.
2. **confirmed** : said that something is true.

'That's easy,' Holmes replied. 'Only three members of the crew are Americans – Captain Calhoun and two others. I also know that the three Americans left the ship last night. I spoke to one of the sailors on the boat, you see. The mail boat is faster than the *Lone Star*. My letter will be waiting for these three men when they arrive – and so will the American police!' he concluded.

Holmes was wrong, however. The murderers of John Openshaw never received the five orange pips that he sent them. The police never arrested them, either. The weather that year was very bad and there was a great storm in the Atlantic Ocean. The *Lone Star* was caught in the storm and she sank [1] without survivors! [2]

---

1. **sank** : went to the bottom of the sea.
2. **survivors** : people that continue to live after a dangerous time.

Go back to the text

1 **Turn the following statements into *What, When, Where, Why* or *How* questions. Then answer the questions.**

Example: They found something in the American Encyclopaedia.
What did they find in the American Encyclopaedia?
They found what 'KKK' was.

a.   The letters were sent from somewhere.

   ....

b.   There was always a delay between the arrival of the letter and the death for some reason.

   ....

c.   The letters 'KKK' stand for something.

   ....

d.   The 'KKK' came into existence during a certain period.

   ....

e.   The 'KKK' was against certain things.

   ....

f.   The 'KKK' sent oak leaves, melon seeds or orange pips for a specific reason.

   ....

g.   The 'KKK' wanted Uncle Elias' diary back for certain reasons.

   ....

h.   According to the police, John Openshaw died in a certain way.

   ....

i.   Holmes sent something to Captain James Calhoun.

   ....

j.   Holmes found out about James Calhoun in a certain manner.

   ....

k.   The police never arrested James Calhoun for a certain reason.

   ....

'Was he frightened of something?'

Some, any, somebody, anybody (or someone, anyone), **something, anything, somewhere, anywhere** are used in the following ways:

a. If the sentence is positive we use *some* (*somebody, something, somewhere*).

b. If the sentence is negative or a question we use *any* (*anybody, anything, anywhere*).
    But there are several exceptions!

c. If we are offering or giving we say: *'Would you like some tea?'* or *'Would you like to go somewhere this afternoon?'*

d. If we expect or think that the answer to our question will be yes: *'Do you think he has lost something?'*

e. *Any* (*anybody, anything, anywhere*) can also mean 'It is not important which one': *'You can ask anybody and they can tell you where Mr Sherlock Holmes lives.'* = 'It doesn't matter who you ask (in other words, everybody knows) and he/she can tell you where Mr Sherlock Holmes lives.'

2 **Below are sentences from *The Five Orange Pips*. Say which of the cases explained above (a, b, c, d or e) applies.**

1 Holmes was working with some papers.

2 He said you could solve any mystery!

3 He never had any visitors.

4 He did not allow anyone to go in there.

5 The box was empty, but my father found some other papers in the attic.

6 Above the letters 'KKK' there was some writing.

7 Something terrible had happened.

8 Was he frightened of something?

3 **Now fill in the gaps with *some*, *any*, *somebody*, *anybody*, *something*, *anything* or *somewhere*.**

a. There was no sign of .... violence.

b. Mr Holmes, do you think .... killed my father? Or do you really think it was an accident?

c. Do you want .... advice, Mr Openshaw?

d. Mr Openshaw, I advise you to take .... train from Waterloo Station and return home as quickly as possible.

e. Captain, are there .... Americans travelling on the *Lone Star*?

f. Yes, Mr Holmes, there are .... Americans on board. Three of them, I believe.

g. .... sent letters from Pondicherry, Dundee and London.

h. The murderers of John Openshaw didn't receive .... from Sherlock Holmes because their ship didn't arrive in America.

i. We shall find out .... useful if we look in the American Encyclopaedia.

j. Mr Openshaw's uncle did not allow .... to go into the room in the attic.

k. Was Colonel Openshaw frightened of ....?

l. Mr Holmes went out ....

PET

The criminal organisations file

4 Sherlock Holmes has a file on all the major criminal organisations in the world. He fills in a standard form for each one. Using the full text below from the American Encyclopaedia, fill in the form for Holmes.

Ku Klux Klan

A name derived [1] from an imaginary resemblance [2] to the clicking sound made when cocking [3] a rifle. This terrible secret society was formed by some ex-Confederate soldiers in the Southern States after the Civil War and it rapidly formed local branches in different parts of the country, notably in Tennessee, Louisiana, the Carolinas, Georgia and Florida. Its power was used for political purposes, principally for the terrorising of the black voters and the murdering or driving from the country those who did not agree with them. Its crimes were usually preceded [4] by a warning sent to the marked man in some unusual way – a sprig of oak leaves, melon seeds or orange pips. When the victim received one of these warnings, he could either change his opinions or flee the country. If he decided to face the situation, he would always die in some strange and unpredictable manner. The organisation of this society was so perfect that there is hardly a case on record where any man was able to escape death or in which the people responsible for these crimes were captured by the police. For some years this organisation was very strong, in spite of the efforts of the United States' Government.
Eventually, in the year 1869, the society rather suddenly collapsed, although crimes like the ones described above have been reported since that date.

1. **derived** : came.
2. **resemblance** : similarity.
3. **cocking** : setting in the position ready for firing.
4. **preceded** : came or went before something.

MAJOR CRIMINAL
ORGANISATIONS IN THE WORLD

Full name : (1) ....

Origin of name : (2) ....
....

Founders : (3) ....
....

When founded : (4) ....

Reasons for founding it : (5) ....
....

Criminal aims : (6) ....
....
....

Methods : (7) ....
....
....

The success of the police in fighting it : (8) ....
....

Present activities : (9) ....
....

*Charles Dickens(1839),
by Daniel Maclise,
National Portrait Gallery,
London.*

About the Author

Charles Dickens (1812-70) was the son of a clerk in the navy pay office. His father was imprisoned [1] for debt and the young Dickens was forced to obtain work in a blacking warehouse, [2] an experience that haunted [3] him for the rest of his life. He later became an office clerk. He learned shorthand [4] and became a parliamentary reporter for *The Morning Chronicle*. Dickens was a prolific [5] contributor [6] to magazines and periodicals and many of his novels were written for this kind of publication. His literary reputation was established by his inventive comic writing, but he went on to write serious novels that attacked social problems in Victorian England.

His best-known novels include:

*The Posthumous Papers of the Pickwick Club; Oliver Twist;
Nicholas Nickleby; The Old Curiosity Shop; A Christmas Carol;
Dombey and Son; David Copperfield; Hard Times; Little Dorrit;
A Tale of Two Cities; Great Expectations.*

---

1. **imprisoned** : put in prison.
2. **blacking warehouse** : place where black pigment was stored.
3. **haunted** : returned repeatedly to your mind.
4. **shorthand** : method of writing rapidly.
5. **prolific** : producing a lot.
6. **contributor** : a person who writes articles for a magazine or a newspaper.

Hunted Down

INTRODUCTION

Criminals and detection fascinated Dickens. What is particularly interesting about this grim [1] little story is the way that he subverts [2] the convention [3] that the narrator of a story tells the whole truth about the events he is recounting. [4] Sampson does not lie to the reader, but he does withhold [5] crucial [6] facts in order to provide the story with its suspense [7] and interest. Agatha Christie was to follow this hint in one of her most famous novels, where the first-person narrator actually turns out [8] to be the murderer!

---

1. **grim** : very serious, depressing.
2. **subverts** : overturns.
3. **convention** : generally accepted practice.
4. **recounting** : telling.
5. **withhold** : hold back.
6. **crucial** : very important.
7. **suspense** : feeling of tenseness or excitement about what may happen.
8. **turns out** : is shown.

Before you read

PET

1 Listen to the beginning of Part One and then say whether the following statements are true (T) or false (F). Then read the text and correct the false ones.

	T	F
1. The clerk tells Mr Sampson that the name of the customer is Mr Julius Slinkton.	☐	☐
2. Mr Slinkton is from the Middle Temple.	☐	☐
3. Mr Slinkton knows Mr Sampson's name.	☐	☐
4. Mr Slinkton and Mr Sampson are old friends.	☐	☐
5. Mr Slinkton and Mr Sampson talk for some time in the insurance office.	☐	☐
6. Mr Slinkton came to get information for an insurance policy for a friend.	☐	☐
7. Mr Slinkton is certain that his friend will take out a life insurance policy.	☐	☐
8. Mr Slinkton says that the insurance profession has suffered a great financial loss.	☐	☐
9. Mr Slinkton says that Mr Meltham was unhappy in love.	☐	☐
10. Mr Meltham left the insurance business for financial reasons.	☐	☐

PART ONE

Most people have a chance to see exciting events in their lives. I am the Chief Manager of an insurance office. [1]

I, too, have seen exciting things in my thirty years of work.

My office had one wall that was covered in glass. I could see everybody who came into the insurance company. I liked to study the faces of new customers before I spoke to them. I decided what kind of people they were before they said a word to me. I learned to trust my first impression of people.

The story I want to tell is about a man who came into the company one day. I watched him through the glass in my office. He seemed about forty years old and he was very well dressed. He seemed very polite and he appeared to be quite a

---

1. **insurance office** : business that provides cover for its customers against various risks to property or life.

gentleman. He was talking to one of the clerks. Despite his appearance, I disliked this man as soon as I saw him.

Suddenly the man noticed that I was looking at him. He smiled at me through the glass. Then he took some papers from the clerk and left.

A few minutes later I called the clerk into my office.

'Who was that man?' I asked him.

'That was Mr Julius Slinkton, sir,' the clerk told me. 'He's from the Middle Temple.' [1]

'What did he want?' I enquired.

'He wanted one of our insurance forms,' the clerk replied. 'He said that a friend of yours recommended [2] this company.'

'He knew my name then, did he?'

'Oh, yes, Mr Sampson,' the clerk confirmed. 'He knew your name.'

About two weeks later I went to have dinner with a friend of mine. One of the other guests was Mr Julius Slinkton. He was standing near the fire. He noticed me and he asked our host to introduce him to me. Our host quickly brought him over. The three of us began to talk.

'I thought you knew Mr Sampson already,' our host said.

'No,' Mr Slinkton told him. 'I followed your advice. I went into the insurance office, but I didn't speak to Mr Sampson. I didn't want to disturb him.'

'Did you come to the office to take out an insurance policy?' [3] I asked Mr Slinkton politely. 'Was it a life insurance policy?'

---

1. **Middle Temple** : one of the Inns of Court where barristers have their chambers.
2. **recommended** : praised as being good for a purpose.
3. **policy** : terms of a contract of insurance.

PART ONE

'It's not a policy for me,' Mr Slinkton said. 'It's for a friend of mine. He asked me to get the information for him. I don't know whether he will take out the policy. People often change their minds, don't you think, Mr Sampson?'

'Yes,' I replied.

We began to talk about other things.

'Your profession has suffered a great loss,' Mr Slinkton said suddenly.

I did not know what he was talking about.

'A loss?' I asked in surprise. 'What kind of loss, sir – a financial one?'

Mr Slinkton laughed.

'I don't mean a financial loss,' he explained. 'I was referring to Mr Meltham – '

Now I understood what he was talking about.

'Ah, yes, Mr Meltham,' I agreed. 'That was indeed a sad loss. He was the most brilliant man I have ever known in the insurance profession. But did you know Mr Meltham?' I asked.

'I knew his reputation,'[1] Mr Slinkton told me. 'What a sad story it is! A young man like that suddenly gives up his business and retires[2] from the world.'

I have said that I disliked Mr Slinkton when I first saw him in the insurance office. I still disliked him. I did not think he was really sad about Mr Meltham at all. I decided to ask Mr Slinkton some questions. I wanted to find out more about this man.

'Have you heard why Mr Meltham left his business?' I asked.

'I have only heard stories about it,' he said. 'Apparently

---

1. **reputation** : opinion held about someone by people in general.

2. **retires** : leaves and goes to a quiet or private place.

Mr Meltham was unhappy in love.'

'That's not the truth,' I told him. 'The truth is that the lady died.'

'She died, did she?' Mr Slinkton repeated. 'That's terrible – poor Mr Meltham. How very sad for him!'

I still felt that Mr Slinkton was not sincere. There was something false about his expression of sadness.

Then he said to me, 'You are surprised that Mr Meltham's story affects [1] me so strongly. I can see that, Mr Sampson but I, too, have suffered a terrible loss recently. I have two nieces, you see. One of them, a girl of twenty-three, died recently. The other niece is also not well. The world is a very sad place!'

Now I thought I understood Mr Slinkton. He was a sensitive man who had suffered. I was angry with myself for disliking him. I watched him for the rest of the evening and he seemed to be a good man. He talked politely to everybody and everybody seemed to like him. I decided that my first impression of Mr Slinkton was wrong.

I spoke to our host about Mr Slinkton. He told me that he had not known him for very long. He told me that Mr Slinkton had taken his two nieces to Italy for their health. It was there that one of them had died. He had returned to England afterwards with his other niece. Now I felt that I understood Mr Slinkton. I was deeply ashamed [2] of my previous distrust [3] of him.

---

1. **affects** : causes someody to change in a particular way.
2. **ashamed** : feeling shame or guilt about something.
3. **distrust** : lack of trust.

Go back to the text

1 **Complete the following summary of Part One using single words or phrases.**

Mr Sampson works as the **1**.. .
He has worked there for **2**.................................... . In order to form
an opinion of a new customer Mr Sampson always **3**.......................
....
Mr Sampson also trusts **4**... .

At the beginning of this story Mr Sampson sees a new customer
whose name is Mr Julius Slinkton. Right from the start Mr
Sampson **5**.................................... even though he seems to be a
perfect gentleman.

Two weeks later at a party, Mr Sampson meets Mr Slinkton.
Mr Slinkton tells Mr Sampson that he had gone to the insurance
office **6**.. . Then Mr Slinkton tells
Mr Sampson that he is very sad to hear that **7**...............................
.... Mr Sampson, however, thinks that he isn't being
sincere. Mr Slinkton understands this and tells Mr Sampson about
his **8**.......................... . One of them **9**............................. and that is
why he is so sorry about Mr Meltham.

Later, Mr Sampson talks to his host about Mr Slinkton. The host
confirms the story and adds that Mr Slinkton had gone **10**...............
.... but **11**.. .
After hearing this Mr Sampson feels **12**...
.... because he had thought badly of Mr Slinkton.

'He was talking to one of the clerks.'

The continuous tenses are very important in English. Any action
taking place in the present is described with the **present
continuous** tense. The same is true for the past: when we wish to
describe a scene in the past we use the **past continuous**.

2 Look at this picture of Mr Sampson's insurance office and say what the various people in the picture were doing.

| sleep | write | look at | smile | talk | sit / drink |

A. ....

B. ....

C. ....

D. ....

E. ....

F. ....

Before you read

1 Listen to the beginning of Part Two and choose the correct answer (A, B, C or D).

1. How did Mr Sampson feel about Mr Slinkton when he saw him for the second time in the office?

    A ☐ He felt sorry for him because of his niece.

    B ☐ He was happy to see him.

    C ☐ He disliked him.

    D ☐ He was curious.

2. Why did Mr Slinkton return to the insurance office?

    A ☐ He wanted to talk about his niece.

    B ☐ He needed another insurance form.

    C ☐ He wanted to know if his friend had started an insurance policy.

    D ☐ He wanted to start an insurance policy for himself.

3. What was the amount of the policy that Mr Beckwith wanted?

    A ☐ £2,000

    B ☐ £20,000

    C ☐ £200

    D ☐ £3,000

4. Who did Mr Beckwith want to write the reference for him?

    A ☐ Mr Sampson.

    B ☐ Mr Slinkton.

    C ☐ Mr Slinkton's niece.

    D ☐ His father.

5. Who had come to talk to Mr Sampson earlier that day?

   A ☐ Mr Beckwith.

   B ☐ A private visitor.

   C ☐ Mr Slinkton's niece.

   D ☐ Mr Meltham.

6. When did Mr Beckwith's insurance policy begin?

   A ☐ In April.

   B ☐ In March.

   C ☐ In May.

   D ☐ In July.

7. Where did Mr Sampson see Mr Slinkton after this?

   A ☐ In the office again.

   B ☐ At another party.

   C ☐ On the beach at Scarborough.

   D ☐ In the street.

8. Who was with Mr Slinkton at the time?

   A ☐ His niece.

   B ☐ Mr Beckwith.

   C ☐ Mr Meltham.

   D ☐ His sister.

PART TWO

Two days later I was sitting in my office as usual. I saw Mr Slinkton come into the outer [1] office. As soon as I saw him I disliked him again. Mr Slinkton waved cheerfully at me and came into my office.

'I have come back,' he said, 'because I want to find out what my friend has done with the insurance forms. I want to know whether he has sent them back to the company. His family are worried about him, you see. They want him to buy a good insurance policy.'

'Perhaps I can help,' I said. 'What is your friend's name, Mr Slinkton?' I asked him.

'Beckwith,' he told me.

---

1. **outer** : outside.

Hunted Down

I called the clerk into my office. I asked him to find out if a man called Beckwith had started an insurance policy with the company. The clerk searched [1] through his files for a moment and then he brought me some papers.

'Yes, Mr Sampson,' he said. 'We received these forms from Mr Beckwith. He wants a policy for two thousand pounds and he has asked Mr Slinkton to write a reference [2] for him.'

'Me!' cried Mr Slinkton in surprise. He thought for a moment. 'But of course I can do that for him.'

Mr Slinkton sat down in my office and wrote the reference for Mr Beckwith. He left the forms in my office, said goodbye politely and then left.

Mr Slinkton was not my only visitor that day. Very early that morning someone else had come to see me at my house. The visit was a very private one. No one knew anything about it at all.

Mr Beckwith's insurance policy

---

1. **searched** : looked carefully.
2. **reference** : statement about a person's character and ability.

began in March. I did not see Mr Slinkton again for six or seven
months. I went to Scarborough in September and I saw Mr
Slinkton walking on the beach there. It was early evening and he
greeted me warmly.

Mr Slinkton was with a young lady. He introduced me to her,
explaining that she was his niece. Her name was Miss Niner.

I looked at her carefully. I was sorry to see
that Miss Niner did not
look very well at all. As
we walked along the
sand, Mr Slinkton

pointed to some tracks in the sand. He laughed.

'Your shadow [1] has been here again,' he joked to Miss Niner.

'Shadow? What shadow?' I asked.

'My uncle is joking, Mr Sampson,' she explained. 'There is an elderly gentleman here in Scarborough. He travels around in a hand-carriage. [2] I see him so often that my uncle calls him my shadow.'

As she was speaking we saw the old man's hand-carriage come into sight. There was a frail [3] old man inside. As the carriage was passing us, he waved his arm at me. He called to me by name. I went to see what he wanted. I was away from Mr Slinkton and Miss Niner for about five minutes.

'My niece is very curious,' Mr Slinkton told me when I rejoined [4] them. 'She wants to know who her shadow is.'

'His name's Major Banks,' I told him. 'He's a very rich man, but a very sick one. He's just been telling me what pleasure you both give him. He says it's obvious that you are very fond of one another.'

'It's true we are very close,' Mr Slinkton said very seriously. 'We are alone, you know – since Margaret died.'

Miss Niner looked sad at her uncle's words. The memory

---

1. **shadow** : *(here)* someone following her.
2. **hand-carriage** : old-fashioned kind of wheelchair for invalids pushed by a servant.
3. **frail** : weak.
4. **rejoined** : came back to.

of her sister was clearly still very painful to her. Suddenly she sat down near a rock on the beach. She was pale. Mr Slinkton walked away from us. He, too, seemed very upset by his memories.

Miss Niner began to tell me about her uncle. She said he was a very good, kind man. She told me that she knew she was going to die soon. She was worried about what would happen to her uncle when she died. I saw the hand-carriage coming back towards us along the sand as she was talking. Suddenly I interrupted her.

'Miss Niner,' I said urgently, [1] 'I have something to tell you. You are in great danger! You must come with me and talk to that man in the hand-carriage. Your life depends on it!'

Miss Niner was very shocked by my words. I walked with her to the hand-carriage before she had time to object. I did not stay there with her for more than two minutes. Within five minutes I saw her walking up the beach with a grey-haired man. He had a slight limp. [2] I knew that she was safe with that man.

I went back to the rock and sat down. Mr Slinkton came back soon afterwards. He was surprised that his niece had gone. We talked for a few minutes. He told me that Miss Niner was very ill and he looked sad while he told me. I replied politely to everything he said, but I was holding a

---

1. **urgently** : worriedly and quickly.
2. **had a slight limp** : walked with difficulty because one leg was hurt.

weapon [1] in my pocket as we walked along together.

'Mr Sampson, may I ask you something?' he suddenly enquired. [2] 'What is the news of that poor man Meltham? Is he dead yet?'

'No,' I told him, 'he's not dead yet. But he won't live long, I'm afraid.'

'What a sad place the world is!' Mr Slinkton sighed [3] quietly.

---

1. **weapon** : an object, such as a pistol, used to hurt or kill people.
2. **enquired** : asked for information.
3. **sighed** : breathed out.

Go back to the text

1 **Answer the following questions.**

   a. Who is Miss Niner?

   b. Who is the shadow?

   c. What does Miss Niner say about Mr Slinkton?

   d. Why does Miss Niner go away with the man with the limp?

   e. What does Mr Slinkton tell Mr Sampson about his niece?

   f. What does Mr Sampson hold in his pocket while he is talking to Mr Slinkton?

   g. What does Mr Slinkton want to know about Mr Meltham?

   h. What is Mr Slinkton's reaction to Mr Sampson's answer?

'But he won't live long, I'm afraid.'

To predict the future* in English we can use both the **going to** and the **will** futures.

The **going to** future is used when there is something in the present situation that makes us sure of what is going to happen.

   *Look how fast she is running! She is **going to** win the race.*

   *It's getting really cold. It's **going to** snow.*

Otherwise with the expressions:
- I'm afraid...
- I expect...
- I believe ...
- I'm certain...
- I'm sure...
- I think...
- It is likely...

− probably...

we often use the **will** future. Notice that we don't need to use 'that' in these sentences.

Remember that the contracted form of the negative is 'won't'.

* Remember that when we want to express our intentions for the future we use **going to**, and when we want to show that some future action is an arrangement or a certain plan (an appointment, a holiday we have booked, a date with our girlfriend, etc.) we use the **present continuous**.

2 Make sentences with the elements given using *will*.

    a. I think/Mr Sampson/be/away/for a week.

    b. I'm afraid/his insurance payment/not/be/very high.

    c. I expect/they/not/arrive/on time.

    d. I believe/Mr Sampson/go/to Scarborough.

    e. Mr Beckwith's family/probably/receive £2,000.

    f. Do you think/the story/have/a happy ending?

    g. I'm certain/you/not/like/Mr Slinkton.

    h. He/not/be able to help you/I'm afraid.

How do you think it will all end?

3 Now say how you think the story will end (using the *will* future of course!)

Remember

−   what you read in the introduction about the special technique that Dickens used in writing the story.

−   what Mr Slinkton does for Mr Beckwith.

−   what Mr Sampson tells Miss Niner to do.

−   that Mr Sampson had a special visitor before Mr Slinkton.

Before you read

1 **Read and choose the correct word (A, B, C or D) for each space and then listen to check your answers.**

It was November before I ¹..... Mr Slinkton again, this time in London. I had a very important appointment at Middle Temple. I ²..... at the Temple and went up ³..... stairs. There were two doors at the top of the stairs. The name BECKWITH was painted on one door. The name SLINKTON was painted on the other. I went in the door marked Beckwith. The room was dirty and there were empty bottles ⁴..... . A young man got up when I entered. He walked very unsteadily and he seemed drunk. 'Slinkton's not in yet,' he said loudly. 'I'll ⁵..... .' He went into the corridor and began to shout loudly. 'Hey! Julius! Come in here and have a drink!' he called. Mr Slinkton came into the room. He was very surprised to see me. 'Julius, this is Mr Sampson!' Beckwith ⁶..... us. 'Boil the brandy, Julius!' he said. He gave Mr Slinkton a filthy saucepan. 'Come on, boil the brandy ⁷..... you usually do!' Mr Slinkton was embarrassed at my presence in the room, I could ⁸..... . 'How is your niece, Mr Slinkton?' I asked him quietly. 'I am sorry ⁹..... my niece has left me,' he replied. 'She went away without a word of explanation.' Beckwith held out the saucepan once more. 'Boil the brandy, Julius,' he repeated. 'Give me what you always give me for breakfast, lunch and dinner. Boil the brandy, I tell you!' Now Mr Slinkton looked ¹⁰..... more embarrassed. This was not a pleasant situation for him. He thought for a moment and then he spoke to me.

1.	**A** saw	**B** seen	**C** see	**D** had seen
2.	**A** came	**B** arrived	**C** went	**D** travelled
3.	**A** any	**B** some	**C** a	**D** all
4.	**A** somewhere	**B** anywhere	**C** nowhere	**D** everywhere
5.	**A** call him	**B** call to him	**C** call at him	**D** call by him
6.	**A** offered	**B** told	**C** introduced	**D** gave
7.	**A** the manner	**B** the fashion	**C** the direction	**D** the way
8.	**A** look	**B** watch	**C** see	**D** say
9.	**A** to say	**B** saying	**C** for to say	**D** for saying
10.	**A** also	**B** as well	**C** even	**D** very

PART THREE

It was November before I saw Mr Slinkton again, this time in London. I had a very important appointment [1] at Middle Temple. I arrived at the Temple and went up some stairs. There were two doors at the top of the stairs. The name BECKWITH was painted on one door. The name SLINKTON was painted on the other.

I went in the door marked Beckwith. The room was dirty and there were empty bottles everywhere. A young man got up when I entered. He walked very unsteadily [2] and he seemed drunk.

1. **appointment** : an arrangement to see somebody at a particular time.
2. **unsteadily** : shakily, without complete control.

Hunted Down

'Slinkton's not in yet,' he said loudly. 'I'll call him.'

He went into the corridor and began to shout loudly.

'Hey! Julius! Come in here and have a drink!' he called.

Mr Slinkton came into the room. He was very surprised to see me.

'Julius, this is Mr Sampson!' Beckwith introduced us. 'Boil the brandy, Julius!' he said.

He gave Mr Slinkton a filthy [1] saucepan. [2] 'Come on, boil the brandy the way you usually do!'

Mr Slinkton was embarrassed at my presence in the room, I could see.

'How is your niece, Mr Slinkton?' I asked him quietly.

'I am sorry to say my niece has left me,' he replied. 'She went away without a word of explanation.'

Beckwith held out the saucepan once more.

'Boil the brandy, Julius,' he repeated. 'Give me what you always give me for breakfast, lunch and dinner. Boil the brandy, I tell you!'

Now Mr Slinkton looked even more embarrassed. This was not a pleasant situation for him. He thought for a moment and then he spoke to me.

END

'You're a man of the world, [3] Mr Sampson,' he began. 'I'll tell you the truth.'

---

1. **filthy** : very dirty.

2. **saucepan** :

3. **man of the world** : someone who is not easily shocked.

'No, Mr Slinkton,' I said firmly. [1] 'You'll never tell the truth. I know all about you.'

'You want to save your insurance company some money,' he said calmly. 'You will try to argue that I was responsible for Beckwith's condition – and for his eventual [2] death. But you won't be able to prove that, you know. You won't be able to prove anything!'

Beckwith suddenly picked up his brandy-glass and threw it at Mr Slinkton. The glass cut his forehead [3] and blood began to flow down his face. Mr Slinkton took out his handkerchief and dried his face. As he was doing this, another man came into the room – a man with grey hair who walked with a slight limp. Mr Slinkton looked at this man in surprise.

'Look very carefully at me,' Beckwith cried out. 'You're a rogue, [4] Slinkton, and I've caught you! I took these rooms on purpose, [5] just to catch you. I pretended to be a drunkard [6] in order to catch you and I've done it. You'll never escape now. You see, the last time you went to see Mr Sampson, I had already been to see him myself – I went to his house very early that morning. We know everything. We know what you were planning. You thought you could kill me for the two

---

1. **firmly** : decisively.

2. **eventual** : happening at last.

3 **forehead** :

4. **rogue** : very bad man.

5. **on purpose** : not by accident.

6. **drunkard** : person who is often under the influence of alcohol.

thousand pounds of the insurance policy, didn't you? You wanted to kill me with brandy, didn't you? But you wanted me to die quickly. That's why you also gave me small amounts of poison.' [1]

Mr Slinkton was surprised by Beckwith's behaviour. The young man did not seem at all drunk now. At first Mr Slinkton did not know how to react. Then he found his courage. He was very pale, but he looked coldly at Beckwith. He did not say a word.

'I took these rooms on purpose,' Beckwith went on. 'I knew what kind of man you are, you see. You're the man who's already killed one innocent [2] girl for her money. And now you're slowly killing another one.'

Slinkton laughed.

'Think how stupid you really are!' Beckwith continued. 'You thought I was drinking brandy all day – but I threw most of it away. You never knew that I came into your room at night when you were asleep. I took all your papers, Slinkton. I read your journal, [3] too. It's got all the information about the poisons that you use. It explains everything. I know where the journal is now!'

Slinkton looked at Beckwith questioningly. [4]

---

1. **poison** : a substance which can kill.
2. **innocent** : not having done wrong.
3. **journal** : a written account of what you have done each day.
4. **questioningly** : using a gesture or tone of voice that expresses doubt.

'It's not in your desk,' Beckwith told him.

'Then you're a thief,' Slinkton told him calmly. He spoke calmly, but his face was white.

'I'm your niece's shadow,' Beckwith said quietly.

Suddenly Slinkton lost his calm and his courage. He looked frightened now. Still he said nothing.

'I've watched you all the time,' Beckwith said. 'I knew that you were poisoning Miss Niner. I went to Mr Sampson and told him everything. That man standing at the door is Mr Sampson's servant. The three of us have saved your niece's life!'

Beckwith paused for a moment to look at Slinkton. Then he went on.

'You don't even know my real name,' he said very quietly. 'You asked Mr Sampson several times if he had any news about Meltham. I can give you news about him – I am Meltham!' he announced triumphantly. [1]

'I loved your niece Margaret. I could not save her – but I promised to pursue [2] you to the end. And I've done it!' he cried. 'I've hunted you down, Slinkton.'

Slinkton now looked in horror at the man who was accusing him. He was unable to speak for fear.

'You never knew my real name,' Meltham told him. 'You are seeing me under my real name now for the first time. You will see me again when you answer the charge [3] of murder in

---

1. **triumphantly** : victoriously.

2. **pursue** : find and chase.

3. **charge** : a statement that somebody has done something wrong.

court. And I hope you see me in your imagination – when they put the rope around your neck and the crowd cries out for your death!'

Slinkton turned quickly away from us for a second and put his hand to his mouth. The room suddenly filled with the smell of some chemical. Slinkton gasped, [1] ran a few steps and fell to the floor. He was dead.

Meltham and I made sure that Slinkton was dead. Then we left the room together.

'I have done what I promised to do,' Meltham said sadly to me. 'My life is ended now.'

I did everything that I could to help him, but the poor man died a few months later.

---

1. **gasped** : breathed loudly.

Go back to the text

1 In Part Three we discover that the narrator, Mr Sampson, did not tell us the truth, that Mr Slinkton did not tell the truth and that Mr Beckwith did not tell the truth!

Read the following quotes by these three characters and say which ones are REALLY true and which ones are lies. Correct the ones that are lies and say why the character involved told that lie.

	TOLD THE TRUTH	LIED
PART ONE		
1. Mr Slinkton said, 'I have only heard stories about it. Apparently Mr Meltham was unhappy in love.'	☐	☐
2. Mr Slinkton said, 'I have two nieces, you see. One of them, a girl of twenty-three, died recently.'	☐	☐
3. Mr Sampson said, 'Now I felt that I understood Mr Slinkton. I was deeply ashamed of my previous distrust.'	☐	☐
PART TWO		
4. Mr Slinkton said, 'I have come back because I want to find out what my friend has done with the insurance forms.'	☐	☐
5. Mr Sampson said, 'Mr Slinkton was not my only visitor that day. Very early that morning someone else had come to see me at my house.'	☐	☐
6. Mr Sampson said, 'Miss Niner, I have something to tell you. You are in great danger!'	☐	☐
PART THREE		
7. Mr Beckwith said, 'Come on, boil the brandy the way you usually do!'	☐	☐
8. Mr Slinkton said, 'I am sorry to say my niece has left me.'	☐	☐

2 Now answer the following questions.

a. When does Mr Sampson discover the truth about Mr Slinkton? Does he tell us that he has discovered the truth at that point in the story?

b. What is Mr Beckwith's real name?

c. How did Mr Beckwith trick Mr Slinkton?

PET

The whole truth and nothing but the truth!

3 Before he died, Mr Meltham wrote a complete statement for the police.

Pretend you are Mr Meltham and write your story.

Include the following information:
– where you met Margaret Niner and how you fell in love with her (you must invent this)
– how you discovered that Mr Slinkton wanted to kill her (you must invent this too)
– how you solemnly promised to hunt down Mr Slinkton
– your plan to trap Mr Slinkton
– Mr Sampson's part in the plan
– the conclusion.

Finish your statement for the police.

To the London Police:

I wish to reveal all the facts concerning the murder of Miss Margaret Niner by Mr Julius Slinkton. Mr Slinkton used his knowledge of insurance and poison to kill Miss Niner and I am afraid I, too, am going to die because of his poisoning.

It all began five years ago when I first met Miss Niner. It was love at first sight ...

Mysterious words

4 Complete the following crossword puzzle.

ACROSS

1. An object used to hurt or kill people.
3. This person gives the party.
5. This person comes to the party.
8. Mr Sampson would say, 'Never .... your first impression of a person!' And this story proves his point.
9. Not strong.
12. The person who directs a company.
15. Mr Slinkton will be accused of .... .
17. This drink is good with a cigar, but don't drink too much or it becomes the word in Two Down.
18. If you breathe out heavily because you are sad or contented, you .... .
19. Mr Slinkton killed one of these and then he tried to kill the other one.
20. A little pot generally used for heating up liquids.

DOWN

2. A chemical substance that kills you.
4. If you believe in someone, you .... him.
6. Your mother's or father's brother is your .... .
7. Embarrassed, guilty.
10. A piece of paper with printed questions or spaces for information.
11. A legal accusation.
13. A suggestion or suggestions that you give somebody to help him.
14. If you drink too much of Seventeen Across you will get .... .
16. A bad, nasty person.

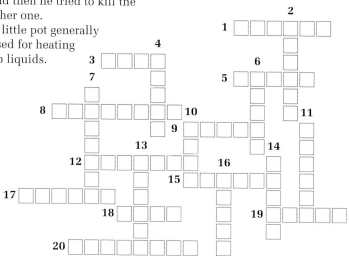

English Detective Fiction

E nglish detective fiction owes a great deal to [1] the American, Edgar Allan Poe. His detective stories were published in the 1840's. Poe's detective character, Dupin, solves crimes by the relentless [2] application of logic. Even when the conclusions that he reaches seem absurd, [3] Dupin is shown to be correct.

Charles Dickens and Wilkie Collins also experimented with detective characters in their novels *Bleak House* (1853) and *The Moonstone* (1868). But it was not until 1887 with the publication of Arthur Conan Doyle's *A Study in Scarlet*, that the genre [4] was really established in English fiction. Conan Doyle's detective, Sherlock Holmes, has many features in common with Poe's detective, Dupin. Both men are unemotional, [5] solitary [6] and intellectual.

Portrait of Edgar Allan Poe, the Mansell Collection.

---

1. **owes a great deal to** : is much influenced by.

2. **relentless** : not stopping or changing.

3. **absurd** : so silly that it makes you laugh.

4. **genre** : particular style or kind of literature.

5. **unemotional** : that is not easily excited.

6. **solitary** : alone.

Where Dupin relies [1] almost exclusively on logic to solve crimes, Holmes uses both logic and scientific observation. The Sherlock Holmes stories were very popular in late Victorian England.

Early in the 20th century the detective story began to change. Full-length detective novels began to replace short stories. This period of the genre is sometimes called 'the Golden Age'. The dominant [2] writers were Agatha Christie, Dorothy Sayers and Margery Allingham.

Agatha Christie created two detective figures: Hercule Poirot and Miss Marple. Hercule Poirot follows the tradition established by Poe and continued by Conan Doyle. He relies on intelligence, 'the little grey cells', to solve crimes. On the other hand Jane Marple, an elderly lady who lives in a picturesque English village, relies on her intuitive [3] understanding of human nature.

Dorothy Sayers created the amateur detective, Lord Peter Wimsey. Her novels are stylish and well-plotted. Margery Allingham's detective is the 'silly ass' [4] Albert Campion. Her novels make use of humour.

Part of the attraction of the 'Golden Age' novels has always been that they give an idealised [5] picture of English life in the first part of the 20th

Agatha Christie, Olympia.

---

1. **relies** : needs for one's support.
2. **dominant** : the most important.
3. **intuitive** : understanding something without reasoning.
4. **'silly ass'** : upper-class fool.
5. **idealised** : imagined as better than reality.

century. They portray the world of the wealthy and leisured [1] upper-middle classes. They are set in grand houses staffed [2] by devoted servants. Most of the characters do not have to work for a living. The much starker [3] realities of the period are not directly explored in the 'Golden Age' novels. The struggle of the Suffragettes, the General Strike of 1926, the appalling [4] poverty of the depression years and the rise of totalitarian [5] regimes in Europe are occasionally mentioned, but Suffragettes, socialists, trade unionists [6] and foreign politicians are marginalised. [7] They are generally treated as shadowy, [8]

Suffragettes on an omnibus advertising a newspaper Votes for Women.

1. **leisured** : having plenty of free time.
2. **staffed** : provided with a group of people working together.
3. **starker** : more serious.
4. **appalling** : terrible.
5. **totalitarian** : having only one political party which controls everything.
6. **trade unionists** : members of a workers' organisation whose purpose is the protection of employees.
7. **marginalised** : made unimportant.
8. **shadowy** : mysterious, difficult to see clearly.

conspiratorial [1] elements that threaten social stability.

The 'Golden Age' writers treated crime as a puzzle that the detective had to solve. This approach to detective fiction changed with the emergence [2] of two important new writers in the 1960's and 1970's. P. D. James and Ruth Rendell broke away [3] from the idealised social settings that had been so popular. They wrote about contemporary England, and struggled for greater realism.

P. D. James created the character Inspector Dalgliesh. Like Dupin and Sherlock Holmes, Dalgliesh is an outsider in society: he is a poet as well as a detective, and he is emotionally detached. [4] Ruth Rendell's novels focus on the psychology of the criminal, as much as the process of crime detection.

1 **Answer the following questions.**

   a. Who created the detective character Dupin?

   b. How does Dupin solve crimes?

   c. How does Sherlock Holmes solve crimes?

   d. Who are Agatha Christie's two famous creations?

   e. How do they solve crimes?

   f. What part of English life did detective writers of the 'Golden Age' describe?

   g. What part of English life did they not describe?

   h. How did P. D. James and Ruth Rendell change the classic British detective novel?

1. **conspiratorial** : secretly planned.
2. **emergence** : appearing.
3. **broke away** : freed yourself from something.
4. **detached** : not involved.

About the Author

Almost nothing is known about the life of **Clarence Rook**, who died in 1915. It is believed that he was an American who settled in London. His best known book is *Hooligan Nights,* which was published in 1899. The book describes life in London's criminal underworld. [1]

---

1. **underworld** : people who are involved in crimes.

The Stir Outside the Cafe Royal

INTRODUCTION

This story shows a new development in the detective genre. In the Sherlock Holmes stories, the emphasis [1] is on the detective's logical reasoning and faultless [2] observation. The emphasis in this story is quite different.

The detective here is a woman and an American. The interest of the tale does not lie in her powers of detection, but in her courage and ingenuity. [3] There is a suggestion that these values are particularly American and modern.

---

1. **emphasis** : special importance given to something.
2. **faultless** : perfect.
3. **ingenuity** : cleverness in making, inventing or arranging things.

Before you read

1 **Listen to the story and choose the correct answer (A, B, C or D).**

1. According to the police Mathurin and a man named Rossiter were
   A ☐ brothers.
   B ☐ two very dangerous criminals.
   C ☐ dead.
   D ☐ the same man.

2. How many people in the world knew the true identity of Mathurin?
   A ☐ Two.
   B ☐ Three.
   C ☐ None.
   D ☐ Four.

3. The horse-drawn cab in which the young lady was travelling went slowly because
   A ☐ the lady said she was frightened of horses.
   B ☐ there was a lot of traffic.
   C ☐ the horse was old.
   D ☐ it was raining.

4. Where did the young lady get out?
   A ☐ At the Café Royal.
   B ☐ At the bank.
   C ☐ At the post office.
   D ☐ At the train station.

5. The girl sat at a table
   A ☐ at the back of the restaurant.
   B ☐ behind the tall, well-dressed man.
   C ☐ at the front of the restaurant.
   D ☐ near the kitchen.

6. What did the young lady ask the waiter to bring her?

   A ☐ Some more water.

   B ☐ Some cake.

   C ☐ Some cheese.

   D ☐ The bill and a sheet of paper.

7. Who did the porter say had stolen the tall man's cigarette case?

   A ☐ The young lady.

   B ☐ A little boy.

   C ☐ A waiter.

   D ☐ The cab driver.

8. Who was the well-dressed man?

   A ☐ The bank manager.

   B ☐ Mathurin, alias Rossiter, alias Connell.

   C ☐ An American detective.

   D ☐ A friend of Mathurin's.

9. Who was the young lady?

   A ☐ A friend of Mathurin's.

   B ☐ A tourist.

   C ☐ An American detective.

   D ☐ Mrs Rossiter.

10. How did Nora Van Snoop recognise Mathurin?

   A ☐ She had seen him shoot her boyfriend, Will Stevens, the bank manager.

   B ☐ She had heard him say his name to a friend.

   C ☐ She had known him since he was a child.

   D ☐ She had seen his photograph.

11. Why did she join the police?

   A ☐ She needed a job.

   B ☐ She was interested in crime.

   C ☐ She wanted to travel to London on important missions.

   D ☐ She wanted to arrest the man who had killed her boyfriend.

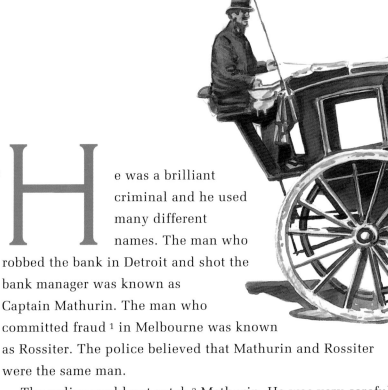

He was a brilliant criminal and he used many different names. The man who robbed the bank in Detroit and shot the bank manager was known as Captain Mathurin. The man who committed fraud [1] in Melbourne was known as Rossiter. The police believed that Mathurin and Rossiter were the same man.

The police could not catch [2] Mathurin. He was very careful to protect his real identity. Most of the people who worked with him did not even know what he looked like.

Only two people in the world could identify [3] him. One of them was the bank manager he had killed in Detroit. Mathurin shot him in front of his girlfriend. It was the other person who ended Mathurin's criminal career.

The Stir Outside the Café Royal

It all happened in a very dull [1] way if you look at it from one
point of view. But the story is very different if you look at it
from another point of view. I first heard the story from a young
detective that I met in a pub near Westminster. Then a young
woman called Miss Van Snoop gave me more information.

A young lady was driving down Regent Street one day in a
horse-drawn cab. [2] It was about one-thirty in the afternoon and
it was warm and sunny. The cab was travelling slowly,
because the young lady said she was frightened of horses.
Regent Street was full of women doing their shopping and
men standing around talking. The young lady looked at the
street with interest.

There was a little stir [3] as the young lady's cab approached [4]
the Café Royal. One cab was stopping outside the restaurant
and there were two others behind it. The traffic in the street
stopped for a moment. The girl looked at the people who were
standing on the steps of the building. She sat back quickly in
her seat.

'Drop me [5] here,' she told the driver. Her accent was
American.

The driver stopped the cab and the girl got out.

She gave the driver a coin. The driver looked at it with
interest. He smiled.

---

1. **dull** : boring.
2. **horse-drawn cab** :
3. **stir** : confusion.
4. **approached** : arrived near.
5. **Drop me** : Let me get out.

The Stir Outside the Café Royal

'Americans are very generous,' he said to himself.

The girl walked towards the Café Royal. She glanced [1] at the men who were standing on the steps. Several of the men looked at her with interest. They were surprised to see a woman on her own. [2] She entered the restaurant and walked into the dining room.

'American, you can be sure of that,' one of the men commented. 'They go anywhere they want. They're not afraid of anything!'

There was a tall man walking in front of the girl

---

1. **glanced** : looked quickly.
2. **on her own** : alone.

towards the dining room. He was very well dressed. He stopped for a moment when he entered the dining room. He was looking for a table. The girl stopped behind him. The waiter waved [1] the man to a table. The girl sat down at a table behind him.

'Excuse me, madam,' the waiter said to her. 'This table is for four people. Would you mind–?' [2]

'I guess [3] I'll stay where I am,' the girl said softly. [4] She gave the waiter a determined look and put some money into his hand.

The restaurant was full of people. Many people looked at the girl who was eating alone, but she did not seem to be embarrassed or shy. She did not look at anyone. When she was not looking at her plate, she kept her eyes fixed on the back of the man at the next table. He ordered champagne with his lunch. The girl drank water. Suddenly she called a waiter.

'Please bring me a sheet of paper and my bill,' she said quietly.

The waiter came back with a sheet of paper. The girl thought for a few minutes. Then she began to write something. She folded the paper and put it in her purse. [5] Then she paid her bill. [6]

A few minutes later the man at the next table paid his bill as well. The girl put on her gloves and watched the man's

---

1. **waved** : moved his hand from side to side in the air to attract somebody's attention.
2. **mind** : object.
3. **guess** : think.
4. **softly** : quietly.
5. **purse** : handbag.
6. **bill** : a piece of paper that shows how much money you must pay for something.

back. The man got up to leave the dining room. He walked past the girl's table. She turned her face away and looked at a mirror on the wall. Then she, too, got up. She followed the man out of the dining room.

The man stopped on the steps for a moment. The porter [1] was talking to a policeman. He noticed the man and asked him if he wanted a cab.

'Yes, please,' the man replied.

Then the porter noticed the girl. She was standing behind the man. As he turned towards her, he saw that her hand was in the man's pocket. She was stealing something. She pulled her hand back quickly.

'What –!' the man cried out. He turned round to face the girl.

'Is something missing, sir?' the porter asked him.

'My cigarette case,' the man said. 'It's gone.'

'What's this?' said the policeman. He stepped forward. [2]

The porter pointed at the girl.

'That woman has stolen this gentleman's cigarette case,' he said. 'I saw her doing it.'

The man looked at the girl.

'Just give it back,' he said quietly. 'I don't want to make a fuss about it.' [3]

'I haven't got it,' the girl answered. 'I'm not a thief. I never touched your pocket.'

---

1. **porter** : a person whose job is to look after the entrance of a building.

2. **stepped forward** : moved to the front.

3. **make a fuss about it** : involve the police in this matter.

The Stir Outside the Café Royal

'I saw her do it,' the porter said again.

'Right!' said the policeman suddenly. 'You'll have to come with me, young lady. You too, sir,' he said to the well-dressed man. 'We'll take a cab to the police station.'

'I didn't steal anything,' the girl said again.

She got into the cab very calmly when it arrived. The policeman watched her carefully. He did not want her to throw anything out of the window. The well-dressed man sat quietly in the cab, looking out of the window.

When they arrived at the police station the girl denied [1] the crime again.

'We'll have to search her,' the inspector decided. She was taken to a room for an interview with the female searcher.

The girl entered the room of the female searcher. As soon as the door was closed she put her hand in her pocket. She took out the cigarette case and placed it on the table.

'There you are,' the girl said. 'Now,' she went on, 'I want you to look in this pocket. Find my purse and take it out.'

The woman took out the girl's purse.

'Open it,' the girl ordered. 'There's a note inside,' she said. 'Read it, please.'

The woman took out the note the girl had written in the restaurant. It said:

---

1. **denied** : said she had not committed.

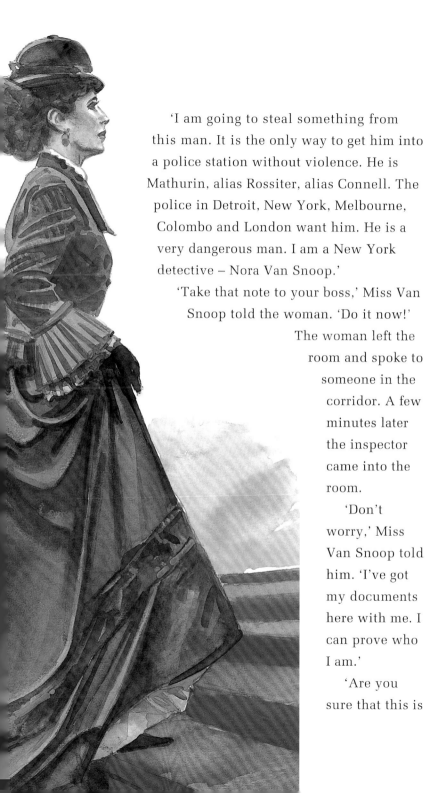

'I am going to steal something from this man. It is the only way to get him into a police station without violence. He is Mathurin, alias Rossiter, alias Connell. The police in Detroit, New York, Melbourne, Colombo and London want him. He is a very dangerous man. I am a New York detective – Nora Van Snoop.'

'Take that note to your boss,' Miss Van Snoop told the woman. 'Do it now!'

The woman left the room and spoke to someone in the corridor. A few minutes later the inspector came into the room.

'Don't worry,' Miss Van Snoop told him. 'I've got my documents here with me. I can prove who I am.'

'Are you sure that this is

the man who shot the Detroit bank manager?' the inspector asked her.

'Heavens!' [1] Miss Van Snoop cried. 'Didn't I see him shoot Will Stevens with my own eyes! Didn't I join the police to find him!'

The inspector left the room. The girl listened attentively. [2] Then she heard a shout from the next room. The inspector came back.

'I think you're right,' he told her. 'It is Mathurin. But why didn't you ask the police to help you?'

'I wanted to arrest him myself,' Miss Van Snoop explained. 'And now I have,' she said quietly. 'Oh, Will! Will!'

Miss Van Snoop sat down and began to cry. Thirty minutes later she left the police station and went into a post office. She telegrammed her resignation [3] from the New York police force. END

---

1. **Heavens!** : exclamation of surprise.
2. **attentively** : carefully.
3. **resignation** : saying that you want to leave your job.

Go back to the text

1 Number the pictures in the correct order. Add the direct quotes
(a-j) and the captions (1-8) to the correct picture.
Not all the direct quotes are actually in the text.

A ☐

CAPTION: ☐

B ☐

CAPTION: ☐

C

CAPTION:

D

CAPTION:

E

CAPTION:

F ☐

CAPTION: ☐

G ☐

CAPTION: ☐

H ☐

CAPTION: ☐

DIRECT QUOTES

a. 'Just give it back. I don't want to make a fuss.'

b. 'Please go slowly because horses frighten me.'

c. 'This table is for four people. Would you mind?'

d. 'I want you to look in this pocket. Find my purse and take it out.'

e. 'I wanted to arrest him myself. And now I have.'

f. 'You'll have to come with me, young lady. You too, sir.'

g. 'I guess I'll stay where I am.'

h. 'But why didn't you ask the police to help you?'

i. 'I'm not a thief.'

j. 'Drop me here.'

CAPTIONS

1. Once inside the police station, the girl admitted that she had stolen the cigarette case and, with the note she had written in the restaurant, she revealed her true identity.

2. The girl easily overcame [1] the waiter's objections. [2]

3. One day a young lady took a cab in search of Mathurin.

4. The young lady saw where Mathurin was going to have lunch and told the cab driver to stop.

5. Only two people in the world could identify Mathurin. One of them was the man he shot, the manager of a Detroit bank.

6. Now we know that Nora Van Snoop had been the bank manager's girlfriend and that she had seen Mathurin shoot him.

7. Outside the restaurant, the porter saw the girl steal Mathurin's cigarette case, but she denied everything.

8. After a little while the girl asked the waiter for the bill and a sheet of paper.

---

1. **overcame** : defeated.   2. **objections** : disapproval.

'The bill, please?'

It is believed that the author of this story was probably American. The heroine [1] is an American, but the story takes place in London.

The writer uses the word 'purse' which in British English is a 'handbag'.

2 **Match the British English words with their American equivalents.** [2]

BRITISH	AMERICAN
to post	lawyer (one who has not taken the bar [3] exam)
draughts	doorman
handbag	check
to think, to suppose	bar
pub	to guess (one use)
bill	to mail
solicitor	checkers
porter	purse

'She was taken to a room for an interview.'

We use the passive in English when we are more interested in indicating the action than the person who does the action.

The **past simple passive** is formed in this way:

Subject + past simple of verb *to be* + past participle of verb.

*She **was taken** to a room.*

If we wish to also indicate who did the action we use 'by'.

*She was taken to a room **by** a policeman.*

---

1. **heroine** : the most important female character in a story.

2. **equivalents** : expressions that have the same meaning.

3. **the bar** : the profession of barrister, who has the right to speak in higher courts.

101

3 Use the elements given to write a sentence using the *past simple passive* and indicate who carries out the action.

Example: bank/rob

The bank was robbed by Mathurin.

a.   Mathurin's criminal career/end.

b.   Mathurin's real identity/know.

c.   The bill and a sheet of paper/bring/to the table.

d.   The note accusing Mathurin/write.

e.   Miss Van Snoop/accuse of stealing the cigarette case.

f.   During the ride to the police station/Miss Van Snoop/watch.

g.   Mathurin/arrest.

h.   Will Stevens/shot.

'They go anywhere they want. They're not afraid of anything.'

4 This story was written more than one hundred years ago. At the time, American women were the only 'liberated' women in the world. For this reason, people were surprised when they saw Miss Van Snoop eat by herself in a restaurant.

Write two or three paragraphs about what young women can do in your country today, and compare it with what their mothers and grandmothers could and could not do.

You can use the expression 'used to', which says what was once true but is no longer so.

Women didn't **use to** *have important positions in companies, but now they do.*

Mothers **used to** *stay home, but now they often work.*

About the Author

Gilbert Keith Chesterton (1874-1936) first became famous as a journalist. He also wrote novels, poetry, literary criticism [1] and a series of essays that took a light-hearted [2] but reflective [3] look at everyday life.

Chesterton's *Father Brown* stories began to appear in 1911. They were very popular with readers, and have retained [4] their popularity. Chesterton's best-known works include:

The Man Who was Thursday: A Nightmare, 1908
The Innocence of Father Brown, 1911

---

1. **criticism** : judgements about the good or bad qualities of literature.
2. **light-hearted** : cheerful.
3. **reflective** : thoughtful.
4. **retained** : continued to have.

THE ORACLE OF THE DOG

INTRODUCTION

In this story Father Brown solves a classic problem in detective fiction. The crime has occurred in a building that has only one entry. [1] Witnesses [2] say that no one went into or came out of the room.

The *Father Brown* stories show a further development in the detective genre. Unlike Sherlock Holmes, Father Brown does not rely on logic or observation to solve crimes. He is interested in the characters of the people involved, and uses his knowledge of human nature to arrive at an intuitive understanding of events.

The *Father Brown* stories are also unusual because they have a strong religious and moral element. In this story, for example, Father Brown argues that the modern world denies the existence of God. The result of this denial [3] is that people have become increasingly [4] superstitious. [5]

---

1. **entry** : *(here)* where you enter a building.
2. **witnesses** : people who are present at the time of a crime.
3. **denial** : saying that something is not true.
4. **increasingly** : more and more.
5. **superstitious** : believing in what cannot be explained by reason or science.

Before you read

1 Fill in the gaps using the words below. Then read the beginning of Part One to see if you were correct.

> article   case   reply   dogs   smiled   lying
> pocket   wonderful   man   excitedly

'Yes,' said Father Brown, 'I like [1]........................ . But I don't like it when people spell them backwards.'

Father Brown was talking to a young [2]........................ . For a moment the young man looked surprised. Then he [3]........................ .

'I see what you mean,' he said. 'You don't like it when people make 'gods' of them. Is that it, Father? But I think dogs are [4]........................ . Sometimes I really do think they know more than we do!'

Father Brown did not [5]........................ . He went on stroking the head of the large black dog that was [6]........................ in front of him.

'Actually,' the young man said [7]........................, 'there's a dog in the [8]........................ I want to talk about. It's a very strange case and the dog is one of the strangest parts of it.'

He reached into his [9]........................ .

'I've got all the details here from the newspaper.'

He passed an [10]........................ from a newspaper to Father Brown.

2 In many books and films animals have supernatural [1] powers or they are as intelligent as or more intelligent than humans. Can you think of an example of a book or film like this?
What do you think? Do animals really have special powers? Or are they just animals?

---

1.   **supernatural** : that cannot be explained by science.

PART ONE

'Yes,' said Father Brown, 'I like dogs. But I don't like it when people spell them backwards.' [1]

Father Brown was talking to a young man. For a moment the young man looked surprised. Then he smiled.

'I see what you mean,' he said. 'You don't like it when people make 'gods' of them. Is that it, Father? But I think dogs are wonderful. Sometimes I really do think they know more than we do!'

Father Brown did not reply. He went on stroking [2] the head of the large black dog that was lying in front of him.

'Actually,' the young man said excitedly, 'there's a dog in the case I want to talk about. It's a very strange case and the

---

1. **backwards** : in the reverse order.
2. **stroking** : moving your hand gently over something.

dog is one of the strangest parts of it.'

He reached into his pocket.

'I've got all the details here from the newspaper.'

He passed an article from a newspaper to Father Brown.

The terrible events at Cranston in Yorkshire are like a mystery story. It is impossible to understand how the murder was committed or what happened to the murder weapon.

Colonel Druce was stabbed [1] from behind while sitting in his summerhouse. [2] There is only one entrance to the summerhouse. There are witnesses who say that no one entered the summerhouse at the time of the murder.

Patrick Floyd, the Colonel's secretary, says that he was working on a ladder in the garden at the time of the murder. He was cutting the hedge [3] with a pair of garden shears. [4] He could see the whole of the garden from that position.

Janet Druce, the Colonel's daughter, was sitting on the terrace at the time of the murder. She confirms that no one entered the summerhouse. She saw Mr Floyd at work in the garden.

Another witness is the Colonel's son, Donald Druce. He

---

1. **stabbed** : attacked with something sharp, like a knife.
2. **summerhouse** : a small building in a garden where people sit during fine weather.
3. **hedge** :
4. **shears** : large scissors used in gardening.

was looking out of his bedroom window at the time of the murder. He says that he saw both Mr Floyd and his sister Janet.

There are two other witnesses – Dr Valentine, and the Colonel's solicitor, Mr Aubrey Traill. Their accounts are consistent with [1] those of the others.

Everybody agrees about what happened. At about 3.30 p.m. Janet Druce went into the summerhouse to ask her father if he wanted some tea. He said he did not want any. He said he was waiting to see his solicitor, Mr Traill. Janet Druce came away from the summerhouse and met Mr Traill on the garden path. He then went into the summerhouse. He stayed there for half an hour. The witnesses saw the Colonel say goodbye to the solicitor. Witnesses say that the Colonel seemed happy that day. He had been angry with his son, Donald, for staying out late the night before. Later, however, the Colonel seemed happy. He had given a very warm welcome to two of his nephews who came over for the day. These two young men were out walking when the tragedy happened.

Ten minutes after the lawyer left the summerhouse, Janet Druce went there again. She found her father lying on the floor with a stab wound [2] to the back. His white linen coat was covered with blood. The Colonel was dead.

---

1. **are consistent with** : are in agreement with.

2. **wound** : injury.

PART ONE

Father Brown studied the newspaper report for a few minutes. Then he put the paper down.

'So the Colonel wore a white coat, did he?' he commented.

'That's right,' Fiennes replied. 'It's a habit he picked up [1] when he lived in the tropics. I didn't see anything myself,' he said. 'I was walking with the two nephews when the murder happened. We had that dog with us – the one I wanted to tell you about. But I saw the lawyer going down the path towards the summerhouse and Floyd working on his ladder. Floyd couldn't have committed the crime.'

'What do you know about the solicitor?' Father Brown asked quietly.

Fiennes was silent for a moment and then he began to speak very seriously.

'Traill's a peculiar [2] man. He dresses very well, but he's quite nervous. He's always moving his hands about, touching his tie or his tie-pin. [3] If I had to say [4] – but it's impossible. No one knows how it was done.'

He stopped talking for a moment. Then he went on.

'But the dog knows who did it. I'm sure of that. That's why I mentioned the murder in the first place. It's the dog that interests me.'

Father Brown did not seem to hear the remark [5] about the

---

1. **picked up** : got into.
2. **peculiar** : strange.
3. **tie-pin** : a pin used to hold a tie in place.
4. **If I had to say** : If you forced me to guess.
5. **remark** : comment.

dog. He wanted more details about the people who were at the house.

'You went to Cranston to visit young Donald Druce, didn't you? Didn't he go on the walk with you?'

'No,' said Fiennes, smiling. 'Donald's a bit wild, [1] you see. He was out all night and only woke in the afternoon. I went on the walk with his cousins, two young army officers from India. The older one, Herbert Druce, talked about horses all the time. The younger one, Harry, talked about his bad luck at Monte Carlo.'

'I see,' said Father Brown. 'Now tell me about the dog,' he asked. 'What kind of dog was it?'

'Like that one,' Fiennes said, pointing to the black dog at Father Brown's feet. 'You said you didn't think it was right to believe in a dog. But I believe in that dog. His name's Nox, by the way. You see, what that dog did is a big mystery – a bigger mystery than the murder itself!'

---

1.   **a bit wild** : *(here)* likes to enjoy himself.

Go back to the text

PET

1 Here is a standard police form for murders. Fill in the details on the form.

FORM NUMBER 8256 —
YORKSHIRE CONSTABULARY [1] HOMICIDE [2] DIVISION

DATE OF CRIME: 4th April 1904 ....

LOCATION: ....

NAME OF VICTIM: ....

ACTIVITY AT MOMENT OF DEATH: ....

IMPORTANT FEATURES OF CLOTHING OR APPEARANCE: ....

....

WITNESSES		
Name	Position at time of crime	What he/she saw
....	....	....
....	....	....
....	....	....

WEAPON:..

WOUNDS: ....

LAST PERSON SEEN WITH VICTIM:...

HOW BODY WAS FOUND: ....

....

1. **constabulary** : the police force of a particular area.

2. **homicide** : murder.

d. Fiennes (walk) .... into Father Brown's study. Father Brown (stand) .... at the window; he (think about) .... superstition in the modern world.

e. When the solicitor (arrive) ...., Donald Druce (sleep) ....

f. We (talk) .... about horses when I (look down) .... and (see) .... a large stick.

g. As we (walk) .... we (pass) .... a curious rock near the sea.

Before you read

PET

1 **Fill in the gaps with the correct words (A, B, C or D). Then read the beginning of Part Two to check your answers.**

Father Brown waited for Fiennes to continue his ¹................. .
'Herbert Druce, his brother Harry and I went for ²................. on the shore with the dog. Colonel Druce's garden is near ³................. .

'As we were walking, we passed a curious rock near the sea. It is called the Rock of Fortune. It's interesting because it's really two rocks, one balanced on top of ⁴................... . It was just as we passed the rock that I had the first feeling. I knew something terrible was ⁵...................!

'I was walking next to Herbert. Harry had stopped to light his pipe under a hedge. We called out to ask him ⁶................... .
Harry looked at his watch and shouted out what time it was.

'We walked on a little further. We were throwing sticks into the sea for Nox, I remember. And then the strange thing with the dog happened. Herbert threw his walking stick into the sea for Nox. The dog jumped into the water and ⁷................... to find the stick.

He came back with it after a **8**.................... . He was very pleased with himself. You know the way dogs are.

'Then Harry threw his walking stick into the sea as well. The dog jumped into the water once more and swam off to find the stick. That's when it happened. Suddenly Nox stopped looking for the stick. He turned around quickly in the sea and swam back to the shore. He stood in front of us and began **9**.................. terribly. It was a fearful noise.

'We all stood and looked at Nox in surprise. The dog howled for a little while and then we heard a woman's shriek. We didn't know what it was then, but we knew afterwards. It was the cry Janet Druce made when she found her father's body in the summerhouse.'

Fiennes paused excitedly.

'So the dog **10**...................., you see! He came out of the sea and began to howl when the Colonel died!'

1.	**A** stories	**B** story	**C** the story	**D** storytime
2.	**A** walk	**B** walking	**C** a walk	**D** walked
3.	**A** sea	**B** in the sea	**C** at the sea	**D** the sea
4.	**A** the other	**B** other	**C** the others	**D** one other
5.	**A** will happen	**B** happened	**C** happens	**D** going to happen
6.	**A** times	**B** what time	**C** about time	**D** the time
7.	**A** swimming off	**B** swim off	**C** swam off	**D** swims off
8.	**A** few minutes	**B** minutes	**C** few minute	**D** fewer minutes
9.	**A** howls	**B** to howl	**C** howl	**D** howled
10.	**A** knew	**B** knowing	**C** known	**D** know

PART TWO

Father Brown waited for Fiennes to continue his story.

'Herbert Druce, his brother Harry and I went for a walk on the shore [1] with the dog. Colonel Druce's garden is near the sea.

'As we were walking, we passed a curious rock near the sea. It is called the Rock of Fortune. It's interesting because it's really two rocks, one balanced on top of the other. It was just as we passed the rock that I had the first feeling. I knew something terrible was going to happen!

'I was walking next to Herbert. Harry had stopped to light his pipe under a hedge. We

---

[1]   **shore** : beach.

called out to ask him the time. Harry looked at his watch and shouted out what time it was.

'We walked on a little further. We were throwing sticks [1] into the sea for Nox, I remember. And then the strange thing with the dog happened. Herbert threw his walking stick into the sea for Nox. The dog jumped into the water and swam off to find the stick. He came back with it after a few minutes. He was very pleased with himself. You know the way dogs are.

'Then Harry threw his walking stick into the sea

---

1. **sticks** : small, thin pieces of wood.

as well. The dog jumped into the water once more and swam off to find the stick. That's when it happened. Suddenly Nox stopped looking for the stick. He turned around quickly in the sea and swam back to the shore. He stood in front of us and began to howl [1] terribly. It was a fearful [2] noise.

'We all stood and looked at Nox in surprise. The dog howled for a little while and then we heard a woman's shriek. [3] We didn't know what it was then, but we knew afterwards. It was the cry Janet Druce made when she found her father's body in the summerhouse.'

Fiennes paused excitedly.

'So the dog knew, you see! He came out of the sea and began to howl when the Colonel died!'

'What happened then?' Father Brown asked calmly.

'We went back to the house,' Fiennes said. 'When we went into the garden we saw the lawyer, Traill,' the young man said. 'Nox saw him too,' he went on. 'The dog jumped forward and barked furiously at the man. Traill turned around and ran away! It was as if the dog knew –'

---

1. **howl** : make a horrible noise.
2. **fearful** : terrible.
3. **shriek** : sudden loud cry.

THE ORACLE OF THE DOG

Suddenly Father Brown stood up. His face was red and he was very angry.

'So the dog knew who the murderer was, did he?' he shouted angrily. 'Traill was accused by the dog, is that what you're saying? What kind of evidence is that, you young fool?' [1]

Fiennes was surprised at the priest's anger. For a moment he was silent.

'What's the matter, Father?' he asked. 'What have I done?'

Father Brown was embarrassed.

'I'm sorry,' he said. 'I've been very rude [2] – please forgive me. Just tell me the rest of the story.'

'You may not believe in the mystery of the dog,' Fiennes said quietly. 'But you've got to admit the dog's behaviour was strange. First he comes out of the sea and begins to howl dreadfully. That happened at the exact moment his master was killed. Then, when he sees the lawyer, the dog barks furiously at him. It must mean something, don't you think?'

The priest said nothing. Fiennes went on.

'Anyway, it wasn't just the dog. There is something odd about Traill. He's a nervous sort of man. He's always playing with that tie-pin of his. That made me think, you see. Of course the police searched everybody as soon as they arrived. They were looking for the weapon, but they never found it. I began to wonder if the tie-pin wasn't the murder weapon.'

---

1. **fool** : someone who is not very clever.

2. **rude** : impolite.

PART TWO

Father Brown nodded his head.

'Ah,' he said thoughtfully, 'the murder weapon. Were there any other suggestions about that?'

'Harry Druce had an idea,' Fiennes told him. 'He was in the Indian police, you see. He knows about detective work. He's a clever young man. He left the police because of some trouble [1] about breaking the rules or something. Anyway, he disagreed with me about the dog. He says a really dangerous dog doesn't bark – it growls.' [2]

'He's right about that,' the priest said softly.

There was silence in the room. Father Brown was thinking. Suddenly he looked up at Fiennes again.

'About the lawyer,' he said. 'Why was Traill at the house that day?'

'He explained about that,' Fiennes said. 'Colonel Druce was preparing a new will.'

'What about the will itself?' Father Brown wanted to know. 'Did the Colonel change the will the afternoon he died?'

'Yes, he did,' Fiennes admitted. 'Druce was angry with his son Donald that day. He changed the will. The money went to Janet rather than Donald.'

'So Janet Druce benefited from her father's death,' Father Brown said quietly.

'Good heavens! You don't mean –?'

---

1. **trouble** : difficulty.

2. **growls** : a dog makes a low noise when it is angry.

THE ORACLE OF THE DOG

'Is she going to marry Dr Valentine?' Father Brown asked.

'I think so,' Fiennes replied. 'They're in love.'

'A doctor always carries his medical bag with him,' Father Brown said quietly. 'There are a lot of sharp objects in a doctor's medical bag, aren't there?'

'You think he could have –?'

Father Brown shook his head.

'The problem is not *who* did it,' he said. 'The problem is *how* they did it. Remember that Colonel Druce was in the summerhouse. There was only one entry to it. A lot of people say that nobody went into the summerhouse. Floyd was on his ladder in the garden. Janet was on the terrace. Donald was looking out of his bedroom window. They all say that no one went into the summerhouse.'

'What do you think?' Fiennes asked Father Brown. 'You've had a lot of experience in this kind of crime.'

'I can't really help,' the priest said. 'I don't know the place or the people. But I'm interested in your friend Harry, the young man from the Indian police. Can you find out what he's doing now?'

'All right,' Fiennes agreed. 'I'm going down to the house again tomorrow. I'll let you know [1] what happens there when I come back.'

---

1. **let you know** : inform you.

Go back to the text

1 **Answer the following questions.**

a. Why was the Rock of Fortune peculiar?

....

b. Were Herbert and Fiennes standing next to Harry when Harry lit his pipe under a hedge?

....

c. What game did they play with Nox?

....

d. When did Nox begin to howl?

....

e. When did they hear Janet's shriek?

....

f. Why did Janet shriek?

....

g. What did Traill do when Nox barked at him?

....

h. Who accused Traill of murdering Colonel Druce?

....

i. Why did Father Brown get angry with Fiennes?

....

j. According to Fiennes, how did Traill murder Colonel Druce?

....

k. Why did Colonel Druce change his will?

....

PET

2 Here are some sentences from Parts One and Two. For each sentence, finish the second sentence so that it means the same as the first.

Example: It is impossible to understand how the murder was committed.
Nobody can understand how the murder was committed.

1. Colonel Druce was stabbed from behind.
   Somebody ....

2. There is only one entrance to the summerhouse.
   The summerhouse ....

3. There are witnesses who say that no one entered the summerhouse at the time of the murder.
   The witnesses say, 'We didn't ....'

4. Janet Druce went into the summerhouse to ask her father if he wanted some tea.
   Janet Druce went into the summerhouse and asked her father, 'Would ....'

5. What that dog did is a big mystery.
   That dog's actions ....

6. Traill was accused by the dog.
   The dog ....

7. He was in the Indian police. He knows about detective work.
   He learned ....

8. He disagreed with me about the dog.
   He said, 'I don't ....'

9. A really dangerous dog doesn't bark – it growls.
   If a dog ....

10. The money went to Janet rather than Donald.
    Janet received ....

'Traill could be the murderer.'

3 You and Father Brown now have many suspects and a lot of evidence. Now it is time to draw some conclusions using the elements given.

Examples:
dog/knows that Colonel Druce was murdered
If you are certain about something write:
The dog can't know that Colonel Druce was murdered because a dog is just an animal!

Donald Druce/be the murderer
If you are only suspicious about something write:
Donald Druce could be the murderer because he knew that his father had changed the will.

a. Traill/the murderer.

b. Tie-pin/murder weapon.

c. Janet Druce/murderer.

d. Dr Valentine's medical bag/the murder weapon.

e. Dr Valentine/murderer.

f. Herbert Druce/murderer.

g. Harry Druce/murderer.

h. Floyd/murderer.

Before you read

1 Think who the murderer is and how he or she committed the crime. Use the elements in the story as the basis for your deduction, but you can invent things too. Just be certain that what you invent does not contradict what you have read!

We think that Dr Valentine killed Colonel Druce, but he did it alone!

....

....

....

....

PART THREE

Two days later Fiennes made another visit to Father Brown's house. The young man entered the room in a state of great excitement. He was very pale.

'You told me to find out what Harry Druce was doing,' he said to Father Brown. 'Do you know what he's done?'

Father Brown did not reply. He looked at his guest calmly.

'I'll tell you what he's done,' Fiennes went on. 'He's killed himself!'

The priest did not seem surprised by the news Fiennes had given him.

'Did you expect this?' Fiennes asked him suspiciously. [1]

'I thought it was possible,' Father Brown replied sadly. 'That young man worried me. That's why I asked you to see what Harry Druce was doing.'

'I found his body,' Fiennes said sadly. 'I was walking

---

1. **suspiciously** : with the feeling that something is wrong.

down the path in the garden when I had a very strange feeling. I knew something was wrong. I looked around me. Everything seemed all right, but I still felt something was terribly wrong. Then I realised what it was. I looked over the top of the hedge. I couldn't see the Rock of Fortune!'

Father Brown listened intently. [1]

'I understood immediately what had happened,' Fiennes went on. 'The top piece of the Rock of Fortune was missing. I ran down the garden and pushed my way through the hedge. I found the loose [2] rock on the shore – and Harry Druce was lying underneath it. There were some words written on the sand near the body: "The Rock of Fortune falls on a Fool." It was a terrible sight.'

'So Harry Druce was the murderer,' Father Brown said sadly. He sat quietly for a moment. 'It was the Colonel's will, you see,' the priest explained. 'Harry Druce thought he was going to inherit everything. He knew the Colonel was angry with your friend Donald. He thought the money would go to him instead.'

He paused for a moment.

'Harry Druce was in trouble, you see. First he lost his job in the Indian police and then he lost his money at Monte Carlo. He murdered his uncle for the money. He killed himself when he discovered that he'd murdered his uncle for nothing.'

'We still don't know how he managed [3] it,' Fiennes said.

---

1. **intently** : with great attention.
2. **loose** : no longer attached.
3. **managed** : succeeded in doing.

'That's what I'd like to know.'

'I think I can tell you how he did it,' the priest offered [1] quietly.

'You!' cried Fiennes excitedly. 'But you haven't been there! You don't know the people. How could you possibly solve the mystery?'

Father Brown jumped up from his chair in excitement.

'The dog!' he shouted. 'The dog, of course! The whole story was right there in front of you, from the start. You didn't look at the dog properly.' [2]

'But you told me that you didn't believe in the dog!' Fiennes argued. 'You got angry when I started telling you about the dog. You were quite rude about it.'

'The dog is the answer to the mystery,' Father Brown repeated. 'But you have to consider the dog as an animal. A dog is not some mysterious force that can judge men. That's where you went wrong.'

The priest paused to think.

'Let me explain what I mean,' he said. 'When you told me about the murder you mentioned the dog's behaviour on the beach and in the garden. You thought the dog 'knew' that the Colonel was dead because it howled dreadfully on the beach. Soon afterwards, Nox barked at the lawyer and you made the same mistake. You thought the dog 'knew' that Traill was the murderer. When Traill ran away from the dog, you were sure he was the murderer.'

---

1. **offered** : said or showed that you were willing to do something.
2. **properly** : correctly.

'But that's what happened!' Fiennes insisted. 'I was there. I saw it.'

'You're very clever with all your psychology,' the priest said. 'That's why you didn't see what was really happening. It was all much simpler than you imagined. Don't you see?' he cried excitedly. 'You made the dog into a kind of mysterious oracle. [1] But a dog isn't an oracle – it's an animal.'

Fiennes looked at the priest in confusion.

'What do you mean?' he asked.

'Traill was a nervous man,' Father Brown said. 'You told me that yourself. You said that he was always playing with his tie-pin, do you remember? Now it's a fact about dogs that they generally don't like nervous people. The dog barked at Traill because he didn't like the man. Traill ran away because he was frightened of the dog. That's all that happened. There was no mystery about it. It had nothing to do with the murder at all.'

Fiennes opened his mouth to speak, but then he changed his mind. Father Brown continued to speak.

'What happened on the beach with the dog was more interesting. Nox jumped into the water to fetch Harry Druce's stick. Then he came back again without it and began to howl. When you learned about the murder, you made the same mistake again. You thought the dog 'knew' the Colonel was dead.'

'But that's what happened,' Fiennes insisted. 'Nox went into the water and he came back without the stick. Then he began to howl. I was there. I saw it.'

---

1. **oracle** : person considered able to give reliable advice.

'There you go [1] again,' Father Brown said impatiently. 'You're treating Nox as if he were some kind of oracle. But a dog is an animal. Dogs don't know anything about murder. They can't detect crime.'

'So what does his behaviour mean, then?' Fiennes asked impatiently.

'Nox went into the water because he wanted to find the stick,' Father Brown explained slowly. 'But he came out of the water for a very simple reason. He came out because he couldn't find the stick. The stick wasn't there! That's why he howled.'

'Why couldn't he find the stick?' Fiennes asked. 'What happened to it?'

'It sank,' Father Brown said simply. 'It wasn't an ordinary walking stick, you see. It was a swordstick. [2] It was the murder weapon and Harry Druce got rid of [3] it in a very clever way. That's why he started that game of throwing things for the dog.'

'A swordstick,' Fiennes said slowly. 'I think I begin to understand it now. But how did Harry Druce kill the Colonel with it?'

'We have to remember two things,' Father Brown told him. 'The murder happened in a summerhouse. The other important thing is that the Colonel was wearing a white coat.'

---

1. **There you go** : You're doing the same thing.
2. **swordstick** : a hollow walking stick containing a sword.
3. **got rid of** : became free of something.

'Go on,' Fiennes said.

'Nobody could understand how the murderer did it,' Father Brown explained. 'There is only one entry to the summerhouse. All the witnesses said that no one entered it. They were right. No one went inside. It wasn't necessary for Harry Druce to enter the summerhouse.'

'What do you mean?'

'A summerhouse is not a solid building. It's made of wooden slats. [1] There are gaps [2] between the wooden slats, aren't there? You've just told me that there was a hedge at the back of the summerhouse. You ran through the hedge when you saw that the Rock of Fortune had fallen. A man could easily stand near the hedge and look through it. The Colonel's white coat made him an easy target.' [3]

Father Brown stopped for a moment.

'You told me something important about your walk on the beach,' he said to Fiennes. 'You said that Harry Druce stopped to light his pipe under a hedge. Do you remember?'

'That's how he did it!' Fiennes said. 'He took out the swordstick and stabbed the spot of white he could see through the hedge.' He thought for a moment. 'But it was terribly risky, [4] wasn't it? He couldn't be certain that the Colonel would die. He couldn't be certain that the Colonel would leave him the money. And in fact he was wrong – he didn't get any money.'

'You have to understand the character of the man,' Father

---

1. **wooden slats** : strips of wood.
2. **gaps** : empty spaces between things.
3. **target** : someone who is being attacked.
4. **risky** : dangerous.

Brown explained. 'Harry Druce was a gambler. [1] He liked risks. He took a risk when he was in the Indian police and he lost his job for it. He took another risk at Monte Carlo and he lost his money for it. He was a gambler by nature. [2] When he came to the house that day, he saw that Colonel Druce was angry with his son Donald. He knew that the Colonel was changing his will. Perhaps the Colonel was going to leave him the money! That was the risk he took. He murdered the Colonel because he thought it was a good risk.'

The two men were silent for a while. Then Fiennes spoke again.

'So the dog really was important to the story?'

'Of course Nox was important to the story,' Father Brown agreed. 'The dog couldn't tell you about the stick because he couldn't talk. You invented the dog's story, instead of really thinking about the dog as an animal. You made Nox into a superstition. [3] That's something that people do all the time, you know. The modern world doesn't believe in God and so it invents magic to take His place. That's why I was angry with you before.'

---

1. **gambler** : person who plays a game (usually cards) for money.
2. **nature** : the qualities of a person.
3. **superstition** : a belief that cannot be explained by reason or science.

Go back to the text

The real meaning of it all

1 Below are different elements in this mystery. Say what the final explanation or importance of the element is. If there is a wrong interpretation given, state that too.

Element	Wrong interpretation	Final interpretation
No one entered the summerhouse at the time of the murder.		
Harry Druce stopped to light his pipe.		
The Colonel was wearing a white coat.		
The Colonel was stabbed in the back but no murder weapon was found.		
Traill was always playing with his tie-pin.		
Nox did not return to the shore with Harry's walking stick.		
Harry was a gambler.		
Nox howled when the Colonel was murdered.		
Nox barked at Traill and Traill ran away.		

2 Why do you think the newspapers called this mystery 'The Invisible Murder Case'?

3 How did Father Brown's knowledge of human and canine [1] nature help him to understand Harry Druce's and Nox's actions?

What do you think?

4 Father Brown says superstition rules in our modern scientific age. He says that people no longer use their common sense or good rational scepticism [2] to judge things. Instead they are ready to believe almost anything they see or hear without any proof at all: this is what he calls superstition.

Do you agree? Can you think of any examples you have read about or seen on television such as miracle [3] cures, UFOs, miracle drugs?
What about advertisements? Do they try to persuade and convince us with rational arguments or do they present miraculous and magical claims?

Write three or four paragraphs in which you say why you agree or disagree with Father Brown.

Animals are very literal

5 Father Brown is very fond of dogs but he does not think that they have special powers of perception [4] or that they can even reason like humans.

What do you think? Are animals really so limited? Have you had any experiences with animals that confirm or contradict what Father Brown believes?

Write three or four paragraphs expressing your point of view.

---

1. **canine** : connected with dogs.
2. **scepticism** : a general feeling of doubt about something.
3. **miracle** : a wonderful event that is impossible according to the laws of nature.
4. **perception** : the ability to notice or understand things.

The devil's ear

6 **In another Father Brown story, *The Purple Wig*, the journalist Francis Finn goes to Devonshire to write about the Eyres, a noble family, and the strange story of the devil's ear of the Eyres.**

It seems that four hundred years ago an ancestor [1] of the Eyres' listened at the door through a keyhole and heard some dark secret of King James I. The King made this ancestor of the Eyres' the Duke of Exmoor so that he would not tell the secret. However, as a kind of magic or divine [2] punishment, all the Dukes of Exmoor have had horribly distorted [3] ears.
More recently the previous Duke of Exmoor, who also wore a wig [4] to cover his horrible ears, almost lost all his lands to his lawyer, Mr Green. This lawyer had tricked the Duke. When the lawyer told the Duke that he wanted half of his lands, the Duke hit him on the head with a bottle and made a triangular cut on the lawyer's bald head. The lawyer then said, 'Good, now I will have all your lands!' But apparently the Duke had not yet lost. He said, 'No, you will not have my lands, because if you try to take them, I will take off my wig! And no man can see my head without my wig and live!'
This then was the story and the legend of the family. The present Duke of Exmoor also wore a wig – a purple wig. He also liked to talk about all the horrible crimes and actions that his ancestors had committed. As a result, everyone who lived near this Duke was terrified of him and his horrible ears and nobody tried to find out if the present Duke's story was true or not. Maybe he isn't even a real Duke!
This journalist, however, with the help of Father Brown, discovers the real reason why the present Duke of Exmoor wears a purple wig. In fact, Father Brown tells the Duke of Exmoor to take off his wig so that everybody can see the truth. The Duke says, 'No, I won't take it off. It would be too horrible for you bear!' But the journalist jumps on the Duke and pulls off the wig. What did Father Brown and Francis Finn see when the Duke was without his wig?

---

1. **ancestor** : a person in your family who lived a long time before you.
2. **divine** : connected with God or a god.
3. **distorted** : badly shaped.
4. **wig** : a covering made of real or false hair that you wear on your head.

To find the answers, solve the puzzle. Find the words. They all come from this story.

1. This is where you can read about recent crimes.

2. This is one of the most common reasons why people commit crimes: they want this!

3. The person who sees somebody committing a crime.

4. In this story, this is the building in which the crime was committed.

5. If you saw a ghost or a bloody body, you would make this sound!

6. If a criminal cuts you with a knife or shoots you in the leg with a gun, then you have a .... .

7. Somebody who likes taking risks to win something.

8. If you hit someone with a knife, you .... him.

9. Caress.

10. Horrible.

11. A criminal is often afraid that these people will find him.

12. Knives, pistols, rifles, swords, swordsticks and sometimes medical instruments or maybe even a tie-pin!

13. This character was not a god but was very important in the story.

14. This information from the story could be useful: a dangerous dog doesn't bark, it .... .

15. To say that somebody committed a crime.

16. Every really good crime or horror story needs some of this liquid.

17. The opposite of forwards.

18. Killing is a crime, and the crime of killing is .... .

19. The professional person who helps prepare wills in Britain.

20. The hot part of the world.

21. If someone wants to steal something from a room on the second or third floor of a house, he can use this.

22. A killer.

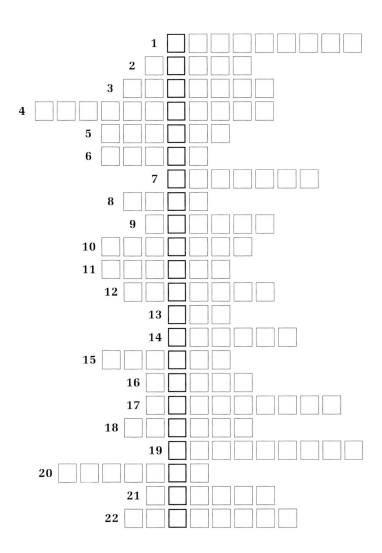

Now answer the following questions. This time, though, you are on your own, and you must be as intelligent as Father Brown himself!

Who is the present Duke of Exmoor?

What is the real reason that he has always worn a purple wig?

EXIT TEST

PET

1 Below are brief descriptions of four people. Decide which of the four stories in this volume would be the most suitable for each person.

1. Bill is a romantic sort. He would do anything for the woman that he loves, and he has! He once travelled all the way from London to San Francisco in California to help his girlfriend who had lost her suitcase and all her money. Bill also really loves his job. He sells cars and he is a great salesman. People who come just to look at a car, go home having bought one, and they don't know how or why!

2. Christine is only eighteen but she is already a professional actress. She loves her job and thinks acting is not a game but something very serious. You can discover very important things about yourself and others by watching a play. 'A play is not real life,' she says, 'but it brings out what is real.'

3. Harry is forty-five years old and lives in a small country town. He doesn't like the modern world much and thinks modern technology is a disaster. He doesn't have a television or a personal computer. He thinks it is much more interesting to go for a walk in the country or sit on a bench in the park to watch people. He says that he can understand everybody's secrets by just watching how they act.

4. Jane is twenty-two years old and goes to university. She is British but she is really interested in American history, especially little-known facts and odd stories. At university she studies chemistry. She loves inventing experiments to discover things.

A ☐ *The Five Orange Pips.*
B ☐ *Hunted Down.*
C ☐ *The Stir Outside the Café Royal.*
D ☐ *The Oracle of the Dog.*

2 A jeweller's was robbed late at night. The door of the shop was still locked and the window was not broken. The owner of the shop is seventy-five years old and he has five sons. Two young women work in the shop.

This is how the crime could have been solved by three different detectives: Miss Van Snoop, Sherlock Holmes and Father Brown. Say which solution belongs to which detective and why you have decided this.

a. *Detective A* had lunch with the owner of the shop. They had sandwiches and tea. As they ate and drank, they talked and talked. The owner told *Detective A* all about his family and the two young women who worked for him. He even told *Detective A* what games his sons liked playing when they were little boys. When the waiter brought the bill for the meal, *Detective A* knew who the thief was. It was the owner's youngest son, who had always broken the other brothers' toys. He had done this because they would never let him play with them. Recently the father and the four other brothers had bought a holiday cottage together without the youngest brother.

....

....

b. *Detective B* asked the owner what had been taken by the thief. The owner told *Detective B* that two very rare and valuable diamonds from South Africa had been stolen, but only a real expert could recognise them. *Detective B* then knew it would be very difficult for the thief to sell them and obtain the right price. *Detective B* looked in the phone directory and discovered that there were three South African jewellers in the town. He went to them and the last one said that he had bought a rare diamond. *Detective B* then asked him to describe the person who had sold the diamond. This was how *Detective B* discovered who the thief was.

....

....

c. *Detective C* simply asked the owner what had been stolen. The owner told *Detective C* that two rare and valuable diamonds from South Africa had been stolen. *Detective C* then went to see a friend who was a journalist for a local newspaper. *Detective C* told the friend to write this article:

RARE SOUTH AFRICAN DIAMONDS NOT STOLEN!

Yes, you have read correctly! Two rare South African diamonds have not been stolen. But why is that news? Well, a local jeweller was clever enough to put excellent copies of those two rare diamonds in his shop. 'I did this,' he explained, 'because I was sure that one day somebody would try to steal those two diamonds! Now,' he continued, 'the real diamonds are in the shop, but I have installed such an excellent alarm system that no thief in the world could possibly steal them!'

That night *Detective C* waited patiently in the shop with the lights off. At 2 a.m., *Detective C* could hear that somebody was turning off the alarm system. At that very moment, *Detective C* called the police and said, 'Come quickly, the thief is entering the shop.'

....

....

The Five Orange Pips

3 **Fill in each gap with one word and then number the paragraphs in the correct order.**

> sank   accident   'KKK'   suicide   owner   America
> unfriendly   fell   ship   police   room   keyhole   storm
> grandfather   revolver   London   five   terrified   letter
> Civil   sundial   sons   vote   pips   papers   orange   War
> coroner   England   organisation   fond   secret   house
> foolish   Klan   never   river   solve

A ☐ So it was that John became the [1]........................ of his
uncle's house about three years before the day he came to
see Holmes. On that day he too received a letter and so
he needed Holmes' help. Besides the terrible letter the
only evidence John had was a piece of blue paper that
came from his uncle's [2]........................ . Holmes advised
him to return home immediately and to place the blue
paper on the sundial. Holmes said that he would stay in
[3]........................ because the secret of the case was there.

B ☐ It all began a long time ago. His [4]........................ had two
[5]........................: John's father and uncle. The uncle went
to [6]........................ when he was a young man and
during the American [7]........................ [8]........................
he fought on the side of the South. After the war he did
not want to give the blacks the [9]........................ .

C ☐ Finally, John's uncle came back to [10]........................ . All
his neighbours said that he was an [11]........................
man, but he seemed to be [12]........................ of John.
When John was about twelve he went to live with his
uncle. Gradually his uncle gave John many
responsibilities in running the house. John could go
where he wanted except for one room in the
[13]........................ . John, of course, looked through the
[14]........................ of the door of this room, but he could
only see some luggage and boxes of [15]........................ .
Everything went fairly well until one day his uncle
received a [16]........................ .

D ☐ Holmes became very angry and promised to catch the
murderer. He discovered the name of the murderer and
that he was returning to America on a [17]........................
called the *Lone Star*. Holmes then sent the murderer,
Captain James Calhoun, a letter containing five orange
pips. This letter would reach Calhoun when he arrived in
America. Holmes also warned the American police that
Calhoun was arriving. But Calhoun [18]........................

received the letter and the police never arrested him because the *Lone Star* 19........................ in the Atlantic Ocean during a great 20........................!

E ☐ After this, John's father went to live in his brother's house, but one day he too received a letter, just like the one John's uncle had received. But John's father did not call the 21........................ . He said they would think he was 22........................ . Instead his father went to visit a friend, but he died while he was there. He 23........................ into a chalk-pit. This time, the coroner said his death was an 24........................ .

F ☐ One day a young man came to Sherlock Holmes for help. His name was John Openshaw. He was very worried because he had received an envelope that contained 25........................ 26........................ 27........................ . He had heard that Holmes could 28........................ any mystery. The young man began to tell his story.

G ☐ On the inside of the flap of the envelope were written the letters 29........................ and inside the envelope were five orange pips. A note said to put some papers on the 30........................ . His uncle looked 31........................ . He brought down the box of papers and burned them. After that his uncle always carried a 32........................ . One day, though, they found his uncle at the edge of a pond in the garden. He was dead. The 33........................ said he had committed 34........................ .

H ☐ When John left, Holmes began his research. He discovered that the three letters 'KKK' stood for Ku Klux 35........................ and that it was a 36........................ 37........................ . Holmes also discovered that the murderer had travelled by ship. Unfortunately, before Holmes could catch the murderer, he and Watson read in the newspaper that John Openshaw had fallen into the 38........................ near Waterloo Bridge. Once again, the police believed it was just an unfortunate accident.

Hunted Down

4 **Answer the following questions.**

a. Why is this story called *Hunted Down*?

....

....

....

b. Mr Sampson says that he always trusted his first impression of people. Does this story show that he was right or wrong in always trusting his first impression?

....

....

....

c. Why is it essential for Mr Sampson to have a good way of judging people?

....

....

....

d. Mr Sampson says that even though he works in an insurance office, he has seen many exciting things. Does this story confirm this? Why or why not?

....

....

....

e. What is so ironic or peculiar about the fact that Mr Slinkton always says, 'It's a very sad world'?

....

....

....

f. How does Mr Slinkton make his money?

....

....

....

....

g. Charles Dickens loved to give his characters names that tell us something about their personality. Look up the verb 'to slink' in your dictionary, and say why Slinkton is a good name for that character.

....
....
....

h. In English there is the expression, 'Fight fire with fire!' It means basically, 'Attack your enemy with the same weapon that your enemy uses to attack you.' How does this expression apply to this story?

....
....
....

The Stir Outside the Café Royal

5 **Answer the following questions.**

a. The author says, 'It all happened in a very dull way if you look at it from one point of view.' What does he mean?

....
....
....

b. The author then says, 'But the story is very different if you look at it from another point of view.' What does he mean?

....
....
....

c. Why is this story called *The Stir Outside the Café Royal*?

....
....
....
....

d. How does this story show that American women were not afraid of anything?

....

....

....

....

e. Look up the word 'snoop' in the dictionary. Is this a good name for Nora?

....

....

....

....

6 Here is Mathurin's point of view. Fill in the gaps with the words given in the box.

> detective   robbed   cab   criminal   champagne
> bank manager   hand   cigarette case   girl   taken
> killed   known   Café Royal   girlfriend   bill

'Yes, I am a great ¹........................ and I have used many names to hide my real name. I have been ²........................ as Mathurin, Connell and Rossiter. I once made a mistake and that mistake is why I am here today in prison.

Several years ago I ³........................ a bank in Detroit. Only two people saw my face: one of them was the ⁴........................ (and I ⁵........................ him!); and the other was his ⁶........................ . But who would have thought that a girl like that would catch me!

However, I never thought of her again. Then one day I decided to have lunch at the ⁷........................ . It was a very normal day for me and I drank ⁸........................ as I always do. Then I paid

the ⁹........................ and went outside to get a ¹⁰........................ .
It was then that I felt a ¹¹........................ in my pocket. I turned
around and saw that ¹²........................! She had stolen my
¹³........................! Of course, I did not want to create a fuss.
However, this girl denied everything and we were
¹⁴........................ to the police station. It was there that this girl
said that she was really a ¹⁵........................ . Not only that, but
she had also been the bank manager's girlfriend!'

The Oracle of the Dog

Oracular Statements

7 **Explain the following oracles in the context of the story.**

a. Sometimes a house is not a house. Why?
....
....
....

b. When is a suit like a target?
....
....
....

c. When is a stick not a stick?
....
....
....

d. My life is devoted to the Good, so it's interesting I am such
an expert in the Bad. Who am I?
....
....
....

e. Sometimes you do better outside than what you usually do inside. Why?

....

....

....

f. This way I am the creator and the king alone. That way I lie under a tree chewing a bone. What am I?

....

....

....

g. Just because dogs don't like me doesn't mean that I am bad. Who am I?

....

....

....

h. This oracle was not an oracle for the person who thought it was an oracle. Why?

....

....

....

i. If you just look for the short, you will not find the long. Why?

....

....

....

j. For me going to get it and bringing it back is great! But not bringing it back is a most horrible fate!

....

....

....

k. He who risks all is soon crushed under Fortune's fall. Why?

....

....

....

Classic

Detective
Stories

KEY T(
THE ACTIVITIE
AND EXIT TES

KEY TO THE ACTIVITIES

THE FIVE ORANGE PIPS

Before you read
Page 11 – Ex. 1
1. A / 2. B / 3. C / 4. C / 5. D / 6. A

PART ONE

Go back to the text
Page 22 – Ex. 1
a. He fought for the Confederacy, i.e., the South.
b. Pieces of luggage and boxes of papers.
c. Five orange pips.
d. Pondicherry, a city in India.
e. Death – his death.
f. 'KKK'.
g. Papers.
h. 'KKK'.
i. His brother, John's father.
j. At the edge of a pond in the garden.
k. He had committed suicide.

Page 23 – Ex. 2
1. D / 2. D / 3. B / 4. A / 5. A / 6. C

Page 24 – Ex. 3
LAW: crime / coroner / will
GAMES: draughts / cards
WEATHER: storm / windy
FAMILY: grandfather / uncle
WORK: factory / retired / client
HOUSES: property / keyhole / landlady / attic / door

```
S O O P G R A N D F A T H E R
T H I S T F O X Q U T A R Y O
P D A N C Q R B Z B T N N M L
A R C Q C A R D S L I W V L D
X A O W L M P L D A C R I A U
V U R P I D P O E W A E U N R
O G O B E L K D F Y E T U D D
L H N O N R L K A E L I P L L
A T E X T F T E C R M R S A A
T S R L I P C Y T V E E T D O
R B I N M Y U H O L P D A Y L
Q U A C P D O O R B R O I C L
S S T O R M I L Y O U W I N D
Q Z B C R I M E O U Q B A R E
```

Before you read
Page 25 – Ex. 1
1. C / 2. B / 3. A / 4. A / 5. D / 6. A / 7. C / 8. C

PART TWO

Go back to the text
Page 33 – Ex. 1
a. F – He didn't call the police because he was afraid they would think he was foolish.
b. F – He went to see his friend about three days later.
c. T
d. F – He had died as a result of an accident.
e. T

f. F – He lived in the house very happily until he too received the five orange pips.

g. T

h. T

Page 33 – Ex. 2

1. f / **2.** c / **3.** h / **4.** a / **5.** e / **6.** i / **7.** b / **8.** g / **9.** d

a. sooner / **b.** carefully / **c.** happily / **d.** danger / **e.** refused / **f.** eagerly / **g.** foolish / **h.** empty / **i.** quickly

Page 35 – Ex. 3

Possible answer:

My dearest Julia!
Something terrible has happened and I feel that I should explain everything to you. After all, we are going to be married soon and you should know everything about your future husband. It all began when my Uncle Elias went to the United States as a young man. He lived in Florida and there he became a wealthy man. When the Civil War began he fought for the South because he did not want the black slaves to be free. After the war he returned to Florida and then he came back to England. This was some years ago. He was then murdered after he received an envelope containing five orange pips. After that, my father came to live in my uncle's house. About a year later he also received five orange pips and then he was murdered.
Only yesterday I, too, received a letter with five orange pips and I was sure that I would die. Fortunately, a friend of mine told me about the great Sherlock Holmes. This man can solve almost any crime, so now I feel better. Do not worry my love! I am sure Holmes will save me.
Your loving husband-to-be,
John

Before you read

Page 36 – Ex. 1

smoked / few / kind / reply / something / letter / about / leave / frightened / still / last / ship / died / after / explain / mean.

PART THREE

Go back to the text

page 45 – Ex. 1

a. Where were the letters sent from?
They were sent from Pondicherry, Dundee and London.

b. Why was there always a delay between the arrival of the letter and the death?
Because the letters travelled on a fast mail boat while the murderers travelled on a slower sailing ship.

c. What do the letters 'KKK' stand for?
They stand for Ku Klux Klan.

d. When did the 'KKK' come into existence?
It came into existence after the Civil War.

e. What was the 'KKK' against?
It was against black people being free and having the right to vote.

f. Why did the 'KKK' send oak leaves, melon seeds or orange pips?
They sent these as a warning to the people they were going to kill if they did not change their ways.

g. Why did the 'KKK' want Uncle Elias' diary back?
Because it contained information about the organisation's members.

h. According to the police, how did John Openshaw die?
He fell into the river by accident.

i. What did Holmes send to Captain James Calhoun?
Five orange pips.

j. How did Holmes find out about James Calhoun?
He checked all the ships that had left from Pondicherry, and discovered one called the *Lone Star*. Only three members of the crew were American: Captain Calhoun and two others.

k. Why didn't the police ever arrest James Calhoun?
Because his ship sank in a storm and never reached the United States.

Page 46 – Ex. 2
1. a / **2.** e / **3.** b / **4.** b / **5.** a / **6.** a / **7.** a / **8.** d

Page 47 – Ex. 3
a. any / **b.** somebody / **c.** some / **d.** any / **e.** any / **f.** some / **g.** Somebody / **h.** anything / **i.** something / **j.** anybody / **k.** something / **l.** somewhere

Page 48 – Ex. 4
Full name: Ku Klux Klan
Origin of name: Resemblance to the sound made when cocking a rifle.
Founders: Ex-Confederate soldiers.
When founded: After the American Civil War.
Reasons for founding it: For political purposes.
Criminal aims: To terrorise black voters and to drive anybody from the country who did not agree with them.
Methods: Sending a sprig of oak leaves, melon seeds or orange pips as a warning to the victim to change his ways or flee the country. Those who did not change their ways were then killed in strange and unpredictable ways.
The success of the police in fighting it: The police had very little success in catching them.

Present activities: Although the 'KKK' collapsed in 1869, typical 'KKK'-style crimes have been reported since then.

HUNTED DOWN

Before you read
Page 52 – Ex. 1
1. T
2. T
3. T
4. F – He knows a friend of Mr Sampson's.
5. F – Mr Slinkton leaves without talking to him.
6. T
7. F – He is not sure whether his friend will take out an insurance policy or not.
8. F – He is referring to Mr Meltham.
9. T
10. F – He left because the woman he loved died.

PART ONE

Go back to the text
Page 57 – Ex. 1
1. Chief Manager of an insurance office
2. thirty years
3. studies their faces before speaking to them
4. his first opinion of people
5. dislikes him
6. to get information for a friend
7. Mr Meltham has left the insurance profession
8. nieces
9. died
10. to Italy for his nieces' health
11. one of them died there
12. ashamed of himself

Page 58 – Ex. 2
A. She is sitting and drinking a glass of water.

B. He is writing.
C. He is looking at the two men talking.
D. He is smiling.
E. They are talking.
F. He is sleeping.

Before you read
Page 59 – Ex. 1
1. C / 2. C / 3. A / 4. B / 5. B / 6. B / 7. C / 8. A

PART TWO

Go back to the text
Page 68 – Ex. 1
a. She is Mr Slinkton's niece.
b. He is Major Banks.
c. She says that he is a good, kind man.
d. Because she is in danger.
e. That she is very ill.
f. A weapon.
g. He wants to know if he is dead yet.
h. He says that the world is a sad place.

Page 69 – Ex. 2
a. I think Mr Sampson will be away for a week.
b. I'm afraid his insurance payment won't be very high.
c. I expect they won't arrive on time.
d. I believe Mr Sampon will go to Scarborough.
e. Mr Beckwith's family will probably receive £2,000.
f. Do you think the story will have a happy ending?
g. I'm certain you won't like Mr Slinkton.
h. He won't be able to help you I'm afraid.

Page 69 – Ex. 3
Open answer.

Before you read
Page 70 – Ex. 1
1. A / 2. B / 3. B / 4. D / 5. A / 6. C / 7. D / 8. C / 9. A / 10. C.

PART THREE

Go back to the text
Page 77 – Ex. 1
1. Lied – Mr Meltham was unhappy because Margaret, Mr Slinkton's niece, had died. Mr Slinkton did not want anyone to know his connection with Mr Meltham.
2. Told the truth.
3. Told the truth.
4. Told the truth.
5. Told the truth.
6. Told the truth.
7. Told the truth.
8. Told the truth.

Page 78 – Ex. 2
a. He discovers the truth when he receives the mysterious visitor. No, he doesn't tell the reader the truth at that point in the story.
b. Mr Meltham.
c. He pretended that he was a drunk and that he was drinking himself to death.

Page 78 – Ex. 3
Possible answer:
To the London Police:
I wish to reveal all the facts concerning the murder of Miss Margaret Niner by Mr Julius Slinkton. Mr Slinkton used his knowledge of insurance and poison to kill Miss Niner and I am afraid I, too, am going to die because of his poisoning.
It all began five years ago when I first met Miss Niner. It was love at first sight when I saw her walking with her

sister in a London park. I talked to her for a long time and she began to tell me about her illness. I had studied chemistry at university so I understood that she was being poisoned. When she told me that her uncle knew all about the insurance business too, I was certain that he was killing her. Unfortunately, her uncle took her to Italy – for her health he said – and she died there. When I heard the news of her death I swore that I would hunt her uncle down!

I had a great idea. I pretended to be a young man named Mr Beckwith who drank too much and who wished to take out a life insurance policy. Mr Slinkton immediately helped me to get the policy and he began to prepare my brandy for me, adding a bit of poison each day. I also contacted Mr Sampson, the manager of an insurance office. He and his servant helped me to trick Mr Slinkton. When Mr Slinkton and his niece were in Scarborough, I disguised myself as an old invalid, and became friends with her. In this way we warned her of her danger and Mr Sampson's servant took her away to safety. Once she was safe, Mr Sampson came to my office at Middle Temple to accuse Mr Slinkton. Unfortunately, when Mr Slinkton saw that we had discovered his horrible plan, he took some poison and died. This then is my story.

page 79 – Ex. 4

1 W E A P O N
3 H O S T
5 G U E S T
8 D I S T R U S T
9 F R A I L
12 M A N A G E R
15 M U R D E R
17 B R A N D Y
18 S I G H
19 N I E C E
20 S A U C E P A N

(crossword grid with numbers 2, 4, 6, 7, 10, 11, 13, 14, 16)

ENGLISH DETECTIVE FICTION

Page 83 – Ex. 1

a. Edgar Allan Poe.
b. By the relentless application of logic.
c. He uses both logic and scientific observation.
d. Hercule Poirot and Miss Marple.
e. Hercule Poirot relies on his intelligence like Dupin and Holmes, while Miss Marple relies on her intuitive understanding of human nature.
f. They described the life of the wealthy and leisured upper-middle classes.
g. They did not describe the lower classes or the social problems of the day.
h. They wrote about all aspects of British society in more realistic terms.

THE STIR OUTSIDE
THE CAFÉ ROYAL

Before you read
Page 86 – Ex. 1
1. D / **2.** A / **3.** A / **4.** A / **5.** B / **6.** D /
7. A / **8.** B / **9.** C / **10.** A / **11.** D

Go back to the text
Page 97 – Ex. 1
A. 8 / Caption 6 / e, h
B. 4 / Caption 2 / c, g
C. 5 / Caption 8
D. 3 / Caption 4 / j
E. 1 / Caption 5
F. 6 / Caption 7 / a, i, f
G. 7 / Caption 1 / d
H. 2 / Caption 3 / b

Page 101 – Ex. 2

BRITISH	AMERICAN
to post	to mail
draughts	checkers
handbag	purse
to think, to suppose	to guess
pub	bar
bill	check
solicitor	lawyer
porter	doorman

Page 102 – Ex. 3
a. Mathurin's criminal career was
ended by Nora Van Snoop.
b. Mathurin's real identity was
known by only two people.
c. The bill and a sheet of paper were
brought to the table by the waiter.
d. The note accusing Mathurin was
written by Nora.
e. Miss Van Snoop was accused of
stealing the cigarette case by the
porter.
f. During the ride to the police
station Miss Van Snoop was
watched by the policeman.
g. Mathurin was arrested by the
inspector.

h. Will Stevens was shot by
Mathurin.

Page 102 – Ex. 4
Open answer.

THE ORACLE OF THE DOG

Before you read
Page 105 – Ex. 1
1. dogs / **2.** man / **3.** smiled /
4. wonderful / **5.** reply / **6.** lying /
7. excitedly / **8.** case / **9.** pocket /
10. article

Page 105 – Ex. 2
Open answer.

PART ONE

Go back to the text
Page 111 – Ex. 1
Date of crime: 4th April 1904
Location: Cranston, Yorkshire.
Name of victim: Colonel Druce
Activity at moment of death: The
Colonel was sitting in his
summerhouse.
**Important features of clothing or
appearance:** He was wearing a white
linen suit.

WITNESSES

Name
Patrick Floyd
Janet Druce
Donald Druce
Position at time of crime
On a ladder in the garden
Sitting on the terrace
Looking out of his bedroom window
What he/she saw
No one entered the summerhouse
No one entered the summerhouse
He saw Janet Druce and Patrick Floyd

Weapon: Unknown
Wounds: Stab wound in back

Last person seen with victim: Mr Aubrey Traill, the Colonel's solicitor.

How body was found: Janet Druce entered the summerhouse and found him.

Possible suspects/Other people in the area at time of crime

Name
Dr Valentine
Aubrey Traill
Herbert Druce
Harry Druce

Relation to victim
His daughter's fiancé
His solicitor
His nephew
His nephew

Job
Doctor
Solicitor
Army officer
Army officer

Page 112 – Ex. 2

a. I was working on a ladder in the garden.
b. I was sitting on the terrace.
c. I was looking out of my bedroom window.
d. I was out walking with Fiennes and Herbert.
e. I was out walking with Herbert and Harry Druce.

Page 113 – Ex. 3

He was wearing a white linen suit.

Page 113 – Ex. 4

a. Harry was telling ... when Janet screamed.
b. I entered ... I saw ... She was ... screaming ... Colonel Druce was lying ...
c. Last night somebody rang ... while I was reading ...
d. Fiennes walked ... Father Brown

was standing ... he was thinking about ...
e. When the solicitor arrived, Donald Druce was sleeping.
f. We were talking ... I looked down and saw ...
g. As we were walking ... we passed ...

Before you read
Page 114 – Ex. 1
1. B / **2.** C / **3.** D / **4.** A / **5.** D / **6.** D / **7.** C / **8.** A / **9.** B / **10.** A

PART TWO

Go back to the text
Page 123 – Ex. 1
a. It was two rocks, one balanced on top of the other.
b. No, they weren't.
c. They were throwing sticks into the water for him to fetch.
d. When Colonel Druce was murdered.
e. Just after Nox began to howl.
f. She had found her father's body.
g. He ran away.
h. Fiennes.
i. Because he said that the dog knew who had killed his master.
j. With his tie-pin.
k. Because he was angry with his son.

Page 124 – Ex. 2
1. Somebody stabbed Colonel Druce from behind.
2. The summerhouse has only one entrance.
3. The witnesses say, 'We didn't see anyone enter the summerhouse at the time of the murder.'
4. Janet Druce went into the summerhouse and asked her father, 'Would you like some tea?'
5. That dog's actions are a big mystery.
6. The dog accused Traill.

7. He learned about detective work in the Indian police.
8. He said, 'I don't agree with you about the dog.'
9. If a dog is really dangerous it doesn't bark, it growls.
10. Janet received the money rather than Donald.

Page 125 – Ex. 3
Possible answers:
a. Traill could be the murderer because Nox barked at him and then he ran away.
b. The tie-pin could be the murder weapon because the police haven't found the actual murder weapon yet, and Traill is a suspect.
c. Janet Druce can't be the murderer because she was on the terrace at the time of the murder.
d. Dr Valentine's medical bag could contain the murder weapon because doctors always carry sharp instruments with them.
e. Dr Valentine could be the murderer because he is going to marry Janet and Janet is going to inherit her father's money now.
f. Herbert Druce can't be the murderer because he was walking with Fiennes at the time of the murder.
g. Harry Druce can't be the murderer because he was walking with Fiennes at the time of the murder.
h. Floyd can't be the murderer because he was working on a

ladder in the garden at the time of the murder.

Before you read
Page 125 – Ex. 1
Open answer.

PART THREE

Go back to the text
Page 133 – Ex. 1

Wrong interpretation	Final interpretation
	Harry Druce stabbed the Colonel with a long swordstick from outside the summerhouse.
	He did this so that Fiennes and Herbert would walk further along and he would have time to stab the Colonel.
	It provided a good target for Harry Druce.
	Harry Druce had used his swordstick, which he then threw in the water.
The tie-pin could have been the murder weapon.	He was simply a nervous man.
He returned to the shore at that moment because he sensed that his master had been murdered.	He returned because he couldn't find the walking stick.
	It was in his nature to take risks. So he risked killing his uncle for his money, even if he wasn't sure that he would inherit it.
The dog knew that he had been murdered and that is why he began to howl.	The dog howled because he couldn't find the walking stick.
Nox knew that he was the murderer and Traill ran away because the dog had discovered him.	Nox barked at him because he was a nervous man and dogs don't like nervous people.

Page 134 – Ex. 2
Because witnesses were certain that

155

nobody entered the summerhouse when the Colonel was killed and no murder weapon was found.

Page 134 – Ex. 3

He understood that Harry was a person who liked taking big risks. It was a big risk to murder his uncle since he was not even sure that he would inherit his money.

He could begin to think correctly about Nox's actions because he knew that a dog could not mysteriously sense a murder. He also knew that dogs do not like nervous people and that would have been enough reason for its barking at Traill.

Page 134 – Ex. 4

Open answer.

Page 134 – Ex. 5

Open answer.

Page 135 – Ex. 6

The present Duke of Exmoor is the lawyer, Mr Green.

The real reason that he has always worn a purple wig is that he has a triangular cut on his bald head!

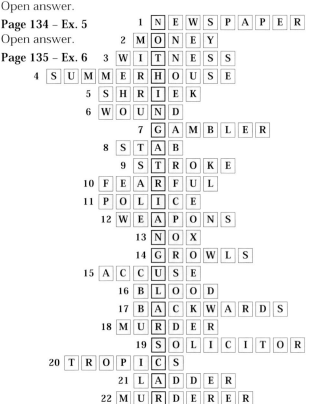

1. NEWSPAPER
2. MONEY
3. WITNESS
4. SUMMERHOUSE
5. SHRIEK
6. WOUND
7. GAMBLER
8. STAB
9. STROKE
10. FEARFUL
11. POLICE
12. WEAPONS
13. NOX
14. GROWLS
15. ACCUSE
16. BLOOD
17. BACKWARDS
18. MURDER
19. SOLICITOR
20. TROPICS
21. LADDER
22. MURDERER

Page 138 – Ex. 1
A. 2 / **B.** 3 / **C.** 4 / **D.** 1

Page 139 – Ex. 2

a. Father Brown is *Detective A* because he relies heavily on his knowledge of human nature to help him interpret the evidence of a crime.

b. Sherlock Holmes is *Detective B* because he uses deduction and then carefully and purposefully-gathered information to help him solve the crime.

c. Miss Van Snoop is *Detective C* because she makes the thief believe that he had worthless diamonds just as she made Mathurin believe that she had stolen his cigarette case.

The Five Orange Pips

Page 140 – Ex. 3
A. 6 / **B.** 2 / **C.** 3 / **D.** 8 / **E.** 5 / **F.** 1 / **G.** 4 / **H.** 7
1. owner / **2.** room / **3.** London / **4.** grandfather / **5.** sons / **6.** America / **7.** Civil / **8.** War / **9.** vote / **10.** England / **11.** unfriendly / **12.** fond / **13.** house / **14.** keyhole / **15.** papers / **16.** letter / **17.** ship / **18.** never / **19.** sank / **20.** storm / **21.** police / **22.** foolish / **23.** fell / **24.** accident / **25.** five / **26.** orange / **27.** pips / **28.** solve / **29.** 'KKK' / **30.** sundial / **31.** terrified / **32.** revolver / **33.** coroner / **34.** suicide / **35.** Klan / **36.** secret / **37.** organisation / **38.** river

Hunted Down

Page 143 – Ex. 4

a. Because Mr Meltham swore that he would 'hunt down' the man who had killed the woman he loved.

b. Yes, in the end his first impression about Mr Slinkton was correct. There is one moment, however, when Slinkton tells him about the death of his niece, that Mr Sampson feels bad about his negative opinion.

c. Because people often try to cheat insurance companies and if Mr Sampson cannot judge the people who take out policies he risks losing large amounts of money.

d. Yes, it certainly does. The story of how he and Meltham trick Slinkton is full of surprises, danger and suspense.

e. It is ironic because he is one of those people who make it a sad world.

f. He poisons people so that he will receive the money from their life insurance policies.

g. 'To move about quietly and secretly to avoid being seen.' Mr Slinkton goes about his job in a secretive fashion because if he is noticed at work, he would be hanged.

h. Mr Meltham pretends to be a drunkard and he also pretends to be the 'shadow'. Mr Sampson also pretends in order to trick Slinkton. These elaborate lies are the same techniques that Slinkton used to trick his victims.

The Sir Outside the Café Royal

Page 144 – Ex. 5

a. Miss Van Snoop's plan required no guns or violence, and she brought Mathurin to the police station in a very 'dull' and calm way.

b. Once you know all the facts (Nora witnessed the murder of her fiancé and joined the police force just to catch the murderer in a clever and

157

daring way) behind the events ('stir' outside the Café Royal, the young American woman alone at the table and her being accused of stealing a cigarette case) then it becomes a very exciting story.

c. The 'stir' refers to the little traffic jam in front of the Café Royal when Mathurin's carriage stops there so that Mathurin can get out and enter the café.

d. Nora not only joined the police force, but she also faced a very dangerous criminal all by herself with complete calm and self-assurance.

e. 'To secretively look into other people's affairs.' Yes, it's a good name for Nora.

Page 145 – Ex. 6
1. criminal / 2. known / 3. robbed / 4. bank manager / 5. killed / 6. girlfriend / 7. Café Royal / 8. champagne / 9. bill / 10. cab / 11. hand / 12. girl / 13. cigarette case / 14. taken / 15. detective

The Oracle of the Dog
Page 146 – Ex. 7
a. The summerhouse in the story does not have solid walls like a real house, and that is how the murderer could kill his victim without actually entering it.

b. The Colonel's white suit made it easier for Harry to stab him.

c. The walking-stick was actually a swordstick and that is why it sunk.

d. Father Brown is a priest.

e. Harry Druce tricked everyone – except Father Brown of course – by killing the Colonel while standing outside the summerhouse. In this way, nobody could understand how the murder took place since nobody entered the summerhouse at the time of the murder.

f. 'God' spelled backwards is 'dog'.

g. Traill, who was suspected by Fiennes because Nox had barked at him.

h. Fiennes thought that Nox was like an oracle, but he did not understand in which way Nox was an oracle. Father Brown understood because he realised that a dog cannot be an oracle in the usual sense of the word. Once Father Brown had eliminated the magical interpretation of the dog's behaviour he could deduce the correct reasons for its barking.

i. No one, except for Father Brown, thought that the murderer could have stabbed the Colonel from outside the summerhouse with a long 'knife', i.e., a sword.

j. Nox the dog, who, as we learn at the end, began to howl because he could not find the stick to bring back to the shore.

k. Harry Druce risked everything by killing his uncle and then he discovered that he had not been named in his uncle's will. This was why he killed himself by having the Rock of Fortune fall on him.

Black Cat English Readers

Level 1A
Peter Pan *CD-ROM*
Zorro!
American Folk Tales
The True Story of Pocahontas
Davy Crockett

Level 1B
Great Expectations
Rip Van Winkle and The Legend of
  Sleepy Hollow
The Happy Prince and The *CD-ROM*
  Selfish Giant
The American West
Halloween Horror

Level 1C
The Adventures of Tom Sawyer *CD-ROM*
The Adventures of Huckleberry Finn
The Wonderful Wizard of Oz *CD-ROM*
The Secret of the Stones
The Wind in the Willows

Level 1D
The Black Arrow
Around the World in *CD-ROM*
  Eighty Days
Little Women
Beauty and the Beast
Black Beauty

Level 2A
Oliver Twist *CD-ROM*
King Authur and his Knights
Oscar Wilde's Short Stories
Robin Hood
British and American Festivities

Level 2B
David Copperfield
Animal Tales
The Fisherman and his Soul
The Call of the Wild
Ghastly Ghosts!

Level 3A
Alice's Adventures in Wonderland
The Jumping Frog
Hamlet
The Secret Garden *CD-ROM*

Great English Monarchs and their
  Times

Level 3B
True Adventure Stories

Level 4A
The £1,000,000 Bank Note
Jane Eyre
Sherlock Holmes Investigates
Gulliver's Travels
The Strange Case of Dr Jekyll and Mr
  Hyde

Level 4B
Romeo and Juliet *CD-ROM*
Treasure Island
The Phantom of the Opera
Classic Detective Stories
Alien at School

Level 5A
A Christmas Carol
The Tragedy of Dr Faustus
Washington Square
A Midsummer Night's Dream
American Horror

Level 5B
Much Ado about Nothing
The Canterbury Tales
Dracula
The Last of the Mohican
The Big Mistake and Other Stories

Level 5C
The Age of Innocence

Level 6A
Pride and Prejudice
Robinson Crusoe
A Tale of Two Cities
Frankenstein
The X-File: Squeeze

Level 6B
Emma
The Scarlet Letter
Tess of the d'Urbervilles
The Murders in the Rue Morgue and
  the Purloined Letter
The Problem of Cell 13

BLACK CAT ENGLISH CLUB

Membership Application Form

BLACK CAT ENGLISH CLUB is for those who love English reading and seek for better English to share and learn with fun together.

Benefits offered:
- *Membership Card*
- *Member badge, poster, bookmark*
- *Book discount coupon*
- *Black Cat English Reward Scheme*
- *English learning e-forum*
- *Surprise gift and more...*

Simply fill out the application form below and fax it back to 2565 1113.

Join Now! It's FREE exclusively for readers who have purchased *Black Cat English Readers* !

The book(or book set) that you have purchased: _____

English Name:_____ (Surname) _____ (Given Name)

Chinese Name: _____

Address: _____

Tel: _____ Fax: _____

Email:_____
Sex: ❏ Male   ❏ Female   (Login password for e-forum will be sent to this email address.)

Education Background: ❏ Primary 1-3   ❏ Primary 4-6   ❏ Junior Secondary Education (F1-3)
❏ Senior Secondary Education (F4-5)   ❏ Matriculation
❏ College   ❏ University or above

Age: ❏ 6 - 9   ❏ 10 - 12   ❏ 13 - 15   ❏ 16 - 18   ❏ 19 - 24   ❏ 25 - 34
❏ 35 - 44   ❏ 45 - 54   ❏ 55 or above

Occupation: ❏ Student   ❏ Teacher   ❏ White Collar   ❏ Blue Collar
❏ Professional   ❏ Manager   ❏ Business Owner   ❏ Housewife
❏ Others (please specify: _____)

As a member, what would you like **BLACK CAT ENGLISH CLUB** to offer:

❏ Member gathering/ party   ❏ English class with native teacher   ❏ English competition
❏ Newsletter   ❏ Online sharing   ❏ Book fair
❏ Book discount   ❏ Others (please specify: _____)

Other suggestions to **BLACK CAT ENGLISH CLUB**:

_____

Please sign here: _____

(Date: _____)